SPACEROCKS

*A collectors' guide to meteorites,
tektites and impactites*

DAVID BRYANT

Spacerocks
A Collectors' Guide To Meteorites, Tektites and Impactities

Copyright © 2018 David Bryant
First paperback edition printed in the United Kingdom
A catalogue record for this book is available from the British Library

ISBN 978-1-9997417-2-3

Published by
Heathland Book

For further copies of this book, please e-mail
info@spacerocksuk.com

Telephone 01603 715933

Designed and typeset by Bob Tibbitts

Printed in Great Britain

Contents

Foreword
by Nik Szymonek

ASTRONOMY is a multi-faceted science that encompasses everything from the origin and evolution of the Universe to galaxies, nebulae, stars and, closer to home, our Solar System of eight planets and all the accompanying material left over from its formation – including the subject of this book, meteorites. Whilst they may not have the glamour of colourful spiral galaxies or mysterious black holes, they have an incredibly powerful story to tell. Virtually all of astronomy relies on analysing the light from unbelievably distant objects and teasing out the underlying physics to understand the complex processes that sculpt our Universe. In this way, astronomy is a passive science.

Meteorites go beyond this as we can hold them in our hands and examine them in unprecedented detail. Imagine being able to touch and interact with a small piece of the Moon or Mars or a distant asteroid. In a slightly more sinister sense you can also hold a fragment of the object that recently exploded over the Russian town of Chelyabinsk, the shockwave causing widespread destruction. It's no wonder that meteorites are highly sought after amongst collectors.

Astronomy popularisers are also multi-faceted personalities. The late Patrick Moore was responsible for introducing more people to the night sky than any other person thanks to his enthusiasm for the subject and friendly, easy-going style of writing. David Bryant shares these desirable characteristics and I can think of no better person to guide you through the fascinating world of meteorites. His friendly and often humorous style belies an impressive scientific background and in this book you'll find sage advice on collecting and displaying meteorite samples along with scholarly information on where they originate from and detailed observations regarding chemical composition. The late Arthur C. Clarke once said that any subject can be made interesting if it's a tale told with enthusiasm and that's certainly the case with 'Spacerocks'. Every meteorite collector should have a copy of it in their collection.

Nik Szymonek, April, 2018

Introduction

LOOKING back, 1961 was a very exciting year to be a 10-year old! Although children in those days were, to a large extent, protected from life's harsher realities such as the Cuba Missile Crisis, two events captured my imagination so completely that they would influence my entire future.

The first of these was the successful launch into Earth-orbit of a Soviet Air Force Major called Yuri Gagarin. Dashing, modest (and, frankly, short!) the familiar shy smile and outsized officer's cap of this unlikely hero were to be seen on every newspaper and magazine cover for weeks after his mission. Gagarin's achievement ultimately resulted in my joining the Royal Navy as a pilot, in the hope I might one day become the UK's first astronaut!

The second, less widely reported event was my finding of what I firmly believed was a meteorite on a family day out in Sussex:

This object was lying in a shallow depression in sand in the grounds of Herstmonceux Castle, then the site of the Greenwich Royal Observatory. When we returned home, my father suggested sending it to Patrick Moore (Back then 'The Sky at Night' was my favourite TV programme: the

Plate 1: My 'meteorite' find

opening music by Sibelius still makes the hair on my neck stand up!) This I duly did, with an accompanying letter explaining why I thought my rock was from outer space. Not only did Sir Patrick reply, he also invited my older brother and me to his home in East Grinstead! Thus began a fascination for astronomy in general – and meteoritics in particular – that has endured for fifty years, ultimately resulting in my wife Linda and I establishing what has become the oldest professional meteorite dealership in the UK.

I have written a number of books (and countless magazine articles!) on a variety of topics, but only one of these concerns itself to any extent with meteorites. At a recent Institute of Astronomy Open Day in Cambridge, a fair number of attendees asked me to rectify this situation, so this is the result!

The intention is to offer a non-specialist guide to meteorites: I'll try to keep the chemistry and physics to an acceptable level and focus on the hobby end of meteoritics.

It's my assumption that the topics that most potential readers would hope to find in this book are:

- How can you recognise a genuine meteorite?
- What are the broad types of meteorites and related material?
- Who can I trust if I want to buy items for my collection?
- How much can I expect to pay?
- How do I display and look after my meteorites?
- Where can I see some meteorites?

Hopefully, you'll find the answers to these questions and many others in this little book!

What are Meteorites?

BROADLY speaking, a meteorite is any piece of solid material that has arrived on Earth from space. Of course, since 1957 we have to add the *caveat* that it must not have come from the Earth in the first place: there are literally millions of pieces of our junk floating around up there, ranging from booster rockets to astronauts' tool-boxes and gloves!

As we'll examine in greater detail in the following chapters, meteorites can be stony, metallic or a mixture of both: it seems increasingly probable that there are even a few meteorites made of ice! Most people are astonished to be told that around 300 tonnes of meteoritic material arrives on our planet **each day!** Of course, the majority of this lands irretrievably in the oceans or undetected in remote regions, or is in the form of dust. When I gave talks to school children, I would often lead meteorite hunts: all that was needed was a plastic bag and a bar magnet. It is the case that the vast majority of meteorites contain iron, and are attracted to a magnet, so using one to search the corners of the playground where windblown dust accumulates generally produced a test-tube full of material in an hour's break-time! Using a hand lens, it is possible to pick out any that are spherical in shape: these are very probably micro-meteorites! (You can conduct a similar search of the sludge in your gutters, after first drying it in the oven and crumbling it into dust.) Another thing I occasionally did with astronomy groups or schools (during a dry spell of weather!) was to peg out a few white cotton bed sheets in a field for a day or two. As before, by gathering the corners together a fair number of spherical micro-meteorites were usually collected.

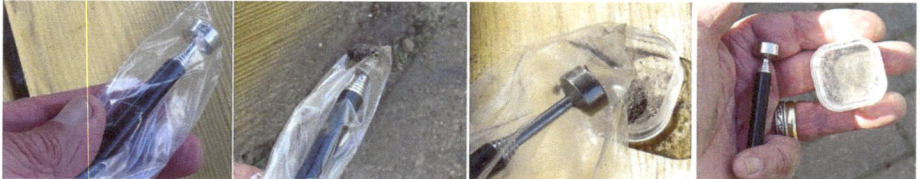

Plate 2: Collecting micrometeorites

That rocks and lumps of iron occasionally fall to Earth from the heavens has been known for thousands of years. In the past, when light pollution was non-existent and people lived much closer to nature, the fall of a meteorite must have been a spectacular and memorable event. Evidence of this (as we'll examine in more detail later) is the fact that meteoric material has been found buried in ritualistic ways in several parts of the world. Perhaps unsurprisingly, the dogma of the Christian Church during the dark ages suppressed this, and it was not until the witnessed fall of the Wold Cottage stone in 1795 that the extraterrestrial origin of meteorites came to be more generally accepted.

Modern meteoritics arguably originates in the work of two men: Harvey Nininger and Eugene Shoemaker.

Nininger was a meteoriticist, teacher and writer who, in 1942, founded the American Meteorite Museum. His collection and study of meteorites reawoke interest among mainstream astronomers: it is widely reported that as late as 1941, Nininger was responsible for identifying half of all the meteorites then known to science! After his death his collection was dispersed: interestingly a large part of it is now owned by the Natural History Museum in London – but you'll have to make an appointment to view it!

Eugene 'Gene' Shoemaker is probably best-remembered for his co-discovery of comet Shoemaker-Levy 9 that spectacularly collided with Jupiter in July, 1994. (The impacts of the several cometary fragments were so violent that my wife Linda and I were able to observe them with just an 80mm refracting telescope) Today, it seems hard for us to believe that as recently as 1960 (following his PhD examination of the Arizona 'Barringer' Crater) Shoemaker was the first person to recognise that it and a number of others were produced by the impact of an extraterrestrial object. Until then it was axiomatic that such structures – even those on the Moon – were volcanic in origin. Shoemaker's careful study of characteristic features such as shocked quartz and shatter cones has resulted in our current understanding of cratering on Earth and on the other rocky members of the Solar System. Shoemaker was an important part of the NASA team that chose the

Apollo landing sites and was also involved in the geological training of the 12 Moonwalkers: had he not been found to be suffering from Addison's Disease, he might well have flown to the Moon aboard Apollo 17.

Before the efforts of these two legendary researchers, meteorites were generally only recognised in the past when they had been seen arriving from space: such objects are referred to as 'falls', as opposed to 'finds', which are discovered by chance lying on the ground. It's obviously easier to retrieve a small lump of burnt rock from a grassy meadow if you saw it land there: of the hundred or so known UK meteorites only half a dozen are finds. (Three of these were located by the **same** meteoriticist and retired dealer, Rob Elliott!) These days, many more people have the knowledge to recognise a meteorite and hunting for them has become big business around the world.

It has become the convention to name a meteorite after the location where it fell or was found: the famous UK Christmas meteorite of 1965, for example, is called Barwell, after the Leicestershire village over which it exploded. However, the vast majority of meteorites in modern collections have been found in the Sahara and exported via dealers in Morocco: they generally are given the catch-all name 'NWA' (north-west Africa) followed by a number reflecting the date of their discovery. At the time of writing the number is in the ten thousands.

Without doubt, many of these numbers refer to the same meteorite, since the tribesmen who are often the original finders like to maximise their incomes by claiming to have made their discoveries in several different locations. Some dealers are not behind the door in this practice, either! In order to have a new and unique find for sale, they'll claim their item is **paired** to an existing, classified specimen: look at online auction sites and see how many lunar meteorites are paired with one another.

In the past (but decreasingly these days, because of the political situation that currently exists across North Africa!) I imported large consignments of totally unclassified meteorites from Morocco, Algeria, Oman and Tunisia. The price was agreed with tribal leaders or local dealers: kind of 'fair trade meteorites'. What you found when you unwrapped a parcel from the Sahara was always a surprise in those days (back in the 1990s) There **would** be meteorites, but also fossils, prehistoric implements, sea shells, pretty minerals and – honestly – dried camel dung on occasions! It was most unusual for the Berber tribesmen to know the exact location of their finds: they didn't have cell phones and GPS devices in those days! And if they did have a rough idea of where they'd made their find, they'd rarely tell you, for fear of western dealers descending on the location. A good example of this is the well-known NWA 869. The large strewn field of this North West African common chondrite was discovered by Berber prospectors in 2001 or 2002 in the region around Tindouf, to the South of Agadir. The exact location has been kept a closely guarded secret by its finders, but over two tonnes of smallish dark grey stones have so far appeared on the market. (The fact that most of the mass of this meteorite is in the form of chestnut-sized individuals suggests that, like the more famous Chelyabinsk stones of 2013, the fall may represent the regolith [outer 'crust'] of a small cometary fragment. More of this later!)

Most of the genuine meteorites exported from North Africa are basic common chondrites (see the next chapter!) which even when cut and polished are fairly unremarkable. These are the 'lowest grade' and most abundant meteorites, but are still fascinating things to hold and reflect upon: after all, they have been touring the Solar System for four and a half billion years and are typical of the building material that made up the planets, asteroids and satellites.

Occasionally something different will turn up: it may be unusually coloured, display large and obvious spherical chondrules, or, unlike the majority of stony meteorites, fail to be attracted to a magnet. In such cases a UK meteoriticist might take or send the specimen to the Meteorite Department at the Natural History Museum in London: in France it would be the 'Université Pierre et Marie Curie' in Paris, in the USA it might be the Dept. of Planetary Sciences at a University: most countries have an accepted authority that can give a definite yea or nay. But beware! In the 'good old days', scientists at these organizations were more than happy to glance at your lump of rock. Times have changed, however! Because of the ever-increasing popularity of meteorite collecting as a hobby, the number of requests for identification has spiraled beyond any possibility of a rapid response. What's more, many meteoriticists have become so jaded by continual demands for verification and classification that they no longer respond. Every year I receive perhaps two hundred e-mails, generally with out-of-focus images attached of someone's precious discovery. I **always** take the time to reply, but it's the case that in over twenty years only once has anyone ever shown me an item that turned out to be a genuine meteorite! I have to say: even I am growing weary of being asked for an off-the-cuff valuation, followed by the inevitable insults when I decline to make one!

However, **genuine** meteorites are eventually classified and catalogued by these various bodies, before appearing on a number of publically-accessible data bases. Probably the most widely consulted is the online **Meteoritical Bulletin** of the Meteoritical Society, which lists every meteorite ever reported to them, together with details of the original finder, total known mass, chemical composition, physical appearance and the organization that classified it.

Generally speaking, then, meteorites are pieces of solid material from space that have reached the Earth's surface after a high-speed passage through the atmosphere. The majority are picked up after an observed fall or are found by knowledgeable searchers: a few turn up quite by chance during building work or other excavations.

The origins and classification of meteorites

PROBABLY the most exciting development in recent years has been the realisation that it is possible to identify the bodies of origin of a good number of meteorites and meteorite 'families'. This has largely been the result of the increased research into their chemical composition and physical structure. In the past it was generally assumed that rocks from space were either 'bits left over from the formation of the Solar System' or 'debris from collisions between two planets': the analysis of meteorites during the last half century has not only refined this model of their origin, but also given planetary astronomers and geologists an insight into the distribution of elements through the primordial solar nebula and the formation and structure of the planets themselves. It is arguably the case that we can infer more about the interior of the Earth from the study of meteorites than has been learned from centuries of geological study!

The basic types of meteorite that have long been recognised are stones, irons and stony irons: let's take a closer look at each.

Stony meteorites are indisputably the most ancient solid material in the Solar System: indeed, some types may include grains that are even older!

Their history begins dramatically billions of years ago with the explosion of a star. As most readers will be aware, it is generally agreed that our Universe began with the spontaneous appearance of energy known as the Big Bang, which is currently believed to have occurred around 13.7 billion years ago. There are literally dozens of models of what went on before, during and after the Big Bang: this basically confirms that cosmologists aren't really certain of the details! In simple terms that even *I* can understand, this seems most likely to have been the sequence of events:

- The entire energy / matter of the present Universe was contained in a tiny singularity: this spontaneously began to **inflate** – and is still doing so.

- During the first second after the Big Bang, energy **uncoupled** into matter, producing increasingly large sub-atomic particles: electrons, positrons, neutrons, protons and the rest.

- The first atoms (hydrogen, followed by helium) formed soon afterwards in a process known as nucleosynthesis.

- Clouds of these gases condensed to form the first **population 111** stars.

- The immense inwards gravitational force within these first generation stars crushed hydrogen and helium atoms together to form larger atoms, up to element 26 in the Periodic Table, Iron.

- This nuclear fusion generated the heat, light, radio waves, x-rays and other electromagnetic radiation that energise our Universe.

- Population 111 stars soon 'used up' their energy, collapsed inwards and exploded as supernovae, spraying metals and other elements heavier than helium into the expanding universe. (Incidentally: stars too large to become supernovae collapsed to form **black holes**)

- Clouds of dust and gas generated by supernovae condensed to form **population 11** stars: recycling of these produced the youngest, metal-rich **population 1** stars (such as the Sun)

- As a star condenses within a **stellar nursery**, the nebula of solid material orbiting it accretes to form larger and larger particles, in most cases eventually becoming a planetary system.

Of particular interest to us as meteorite collectors, is the fact that among the first material to accrete in our solar nebula were the small spherical **chondrules** that make up the major group of stony meteorites, the **chondrites**. The process by which chondrules originally formed remains the subject of debate: some suggestions include shockwaves from a nearby supernova, sudden outbursts of radiation from the Sun, lightning within the Solar Nebula and, more prosaically, heating of stony planessimals during collisions or of meteoroids by ablation as they pass through the Earth's atmosphere. My personal belief is that a combination of the first three might well do the job! The last two seem unlikely, since they fail to account for the various classes of meteorite that don't display chondrules.

Apart from meteorites that originated on a fully-differentiated body (an asteroid, satellite or planet) virtually all stony meteorites contain chondrules: they are found in a wide range of colours and structures and generally measure from .2mm to 1mm. (Although I have personally seen chondrules as much as 20mm in diameter) The largest recorded seem to have been found within the Parnallee chondrite that fell on February 28th, 1857, in Tamil Nadu, India: some of these have measured up to 50mm in diameter!

The largest class of stony meteorites are the **common chondrites** (alternatively 'ordinary chondrites') These all contain varying amounts of nickel-iron, which explains why they are attracted to a magnet! The amount of iron in any meteorite is the basis for its classification:

LL A meteorite containing little metallic iron (< 3%) and around 20% iron in the form of silicates and other minerals. Once called *amphoterites*, these are the rarest of the common chondrites and are often only weakly attracted to a magnet.

L More abundant, these meteorites contain <10% metallic iron as well as minerals such as orthopyroxene, troilite and olivine.

H These are the most abundant of all meteorites, accounting for around 40% of all those known. They contain up to 15% metallic iron and are strongly attracted to a magnet.

The size and abundance of chondrules in a meteorite are reflected in the number that follows its group letter:

3 Containing many, well-defined and large chondrules.

4 Containing less well-defined chondrules: these meteorites show some evidence of chemical weathering or differentiation.

5 Fewer, poorly-defined chondrules present.

6 Chondrules completely absent or very poorly-defined.

Type 3 **unequilibrated** meteorites are considered among the most unaltered and primitive types, the minerals in these chondrites suggesting formations in different regions of the solar nebula. High temperature metamorphosis of Type 4, 5 and 6 chondrites has produced homogenous **equilibrated** compositions with crystallised matrices.

Perhaps a little surprising is the fact that chondrite meteorites all contain hydrated minerals, or even water itself! This tends to suggest that the solar nebula contained ice, which was involved in the process of planetary accretion.

Once temperatures on the parent bodies of the chondrites rose sufficiently for melting to occur, the resulting liquid water must have percolated inwards allowing hydration to occur. It seems likely that both cometary and asteroidal impacts were the sources of a fair percentage of the water that occurred in the past on the larger planets and moons – and on the Earth today!

In addition to the L, LL and H groups of common chondrites, there are several much scarcer classifications, one of which – the carbonaceous chondrites – includes material that is even more ancient than the Solar System!

Let's examine each of these in more detail.

Carbonaceous chondrites are a complex of nine groups, the designation of which are all prefixed by a '**C**' (for carbonaceous!) This is followed by a second letter, which is usually the initial of the type meteorite of the group:

CI chondrite (Ivuna-like) group

CM chondrite (Mighei-like) group

CO chondrite (Ornans-like) group

CV chondrite (Vigarano-like) group

CK chondrite (Karoonda-like) group

CR chondrite (Renazzo-like) group

CH chondrite (Allan Hills 85085-like) group

CB chondrite (Bencubbin-like) group

C chondrites (ungrouped)

As their name implies, the chief characteristic of the group is a dark matrix containing carbon and a 'soup' of organic molecules. Also present are large amounts of magnesium-rich minerals such as olivine and serpentine as well as hydrated minerals and water, especially in the the CI group, which don't appear to have ever been subjected to temperatures above 50°C.

Since many carbonaceous chondrites are unaltered by chemical or physical action they often display the most incredible chondrules. Also found in some carbon-aceous meteorites (particularly the CV group) are whitish Calcium Aluminium Inclusions. Radiometric dating has shown that these were formed before chondrules, with a minimum age of around 4.6 billion years. Their com-position and structure suggest that CAIs were formed at higher temperatures than chondrules. Although the knowledge gained about the formation of the first solid material in the Solar System makes the carbonaceous meteorites scientifically significant, what has captured the popular imagination is the fact that they all contain organic matter, particularly amino acids.

Amino acids are organic (carbon-based) molecules consisting of an amine group, a carboxylic acid group and a variable side chain. Amino acids are often referred to as the 'building blocks' of proteins: these, in turn, are the basic molecules of life itself. On Earth, just twenty amino acids form the basis of all proteins: astonishingly, **over one hundred** have been identified in carbonaceous meteorites! Many researchers theorise that life spontaneously arose on Earth when amino acids arrived in the primordial oceans aboard carbonaceous meteorites four billion years ago. Equally intriguing is the possibility that the same process may well have occurred throughout the universe, suggesting that some form of life may well be found on every planetary body with liquid water on or beneath its surface.

For this reason, carbonaceous meteorites are probably among the most studied and scientifically important of all the chondrites. Naturally, this has resulted in generally higher prices for the most sought-after specimens such as **Murchison** and **Tagish Lake**. Even so, a decent polished slice of, say, **Allende** or **NWA 3118** can still be obtained for around £50. If you add one to your collection, you could introduce it to your friends as not only the oldest object they will ever hold, but also as a genuine alien visitor!

Arguably, the most beautiful members of the class are the twenty or so CBs, named for the type meteorite Bencubbin that was found in Australia in 1930. CBs contain large 'blebs' of nickel iron in sufficient amounts (>50%) that they would be classified as stony irons, were it not for the chemical compostion of their matrix, which is similar to the CRs. Like others in this group, CBs often contained 'armoured chondrules', attractive structures which, in a polished slice, appear as silvery rings with a dark centre.

There is a fascinating story about the Gujba CB (pictured on page 19) which fell in Nigeria in 1984. The original cone-shaped 100kg mass was broken into thousands of small pieces by local villagers, some of which were allegedly swallowed as medicine! (A similar fate befell the Novo-Urei and Mbale meteorites from Russia and Uganda respectively!)

As we move on, chondrite classification becomes a little more fluid, with increasingly scarce types being assigned to small groups, each with just a few members.

The enstatites make up around 2% of the known chondrites: just a couple of hundred meteorites have been given the 'E' classification. The class derives its name from the mineral enstatite, $MgSiO_3$, which is abundant in the two subdivisions of the class: more in **EH** than in **EL**. What these both have in common is an almost total lack of iron oxide, the iron content being in the form of nickel-iron or iron sulphides: for this reason the enstatites are described as the most *reduced* of the chondrite types. This 'oxygen poverty' suggests a formation near the centre of the solar nebula – possibly even within the orbit of Mercury! Members of the group are also among the 'dryest' known meteorites, with less than 0.01% hydrated minerals.

It's often very difficult to see the chondrules in some enstatites, so that some members of the group were originally classified as aubrites (More later!) I have cut and polished literally dozens of slices of NWA 2965 and have only ever found a few small and very indistinct pale-coloured structures that could have been chondrules.

Two other types of chondrite are generally acknowledged: the Rumurutites and the 'K' grouplet.

The two hundred or so R chondrites are named after the type meteorite, Rumuruti, which was seen falling in Kenya in 1934. They are among the most oxidised of the stony meteorites with little or no metallic nickel-iron. Later on we'll look in detail at how measuring the amounts of various atomic forms (isotopes) of oxygen has become a powerful tool in classification and assigning an origin to meteorites: suffice it to say now that the rumurutites have a high concentration of the ^{17}O isotope and appear to originate from the surface layer (regolith) of an asteroid.

There are currently just four members of the **Kakangari** (K) grouplet, all of which have abundant metal (up to 10%) and similar oxygen isotope compositions to the carbonaceous chondrites. The grouplet is named after the original (K3) specimen that was seen falling in India in 1890.

The reader will have noticed that there are no type 7 chondrites: there's a reason for this! Once a planetary body accretes into a large enough object, all kinds of chemical, thermal and physical forces begin to act upon its rocks: tectonic movements, heat generated by radioactive decay, further collisions, freeze-thaw, hydration – all of these are responsible for the metamorphosis of the original chondrite-rich component minerals. The body begins to **differentiate**: heavy elements sink downwards to form a nickel-iron core, while the outer layers of crust and mantle are recycled into new minerals.

Plate 3: Some meteorite slices illustrating various types:

NWA 5205 (L3)
al Haggounia 001 (aubrite)
Gujba (bencubbinite)

Allende (CV3)
Unc. L4 (with armoured
chondrule) NWA 4017 (H4)

Further collisions between these protoplanets and also larger, differentiated asteroidal bodies must have projected millions of tonnes of material back out into the Solar System: it is this debris that occasionally falls to

Earth in the form of achondrite meteorites. There are, at the time of writing, four recognised classes of achondrite, each comprised of several groups:

- **Primitive achondrites** (Acapulcoite/Lodranite group and Winonaites)

- **Asteroidal achondrites** (HEDs, Ureilites, Brachinites, Aubrites and Angrites)

- **Lunar achondrites** (Highland breccias and Maria basalts)

- **Martian achondrites** (Shergottites, Nakhlites, Chassignites, orthopyroxenites and regoliths)

It says much about the rapid advances in the field of meteoritics that it is less than forty years since the existence of planetary meteorites was first accepted, following the discovery of the Antarctic meteorite AH 81005. With the rapid growth in interest in collecting meteorites (and hence the increased discovery of new stones by dealers) new achondrites seem to turn up virtually every week!

Primitive achondrites (PACs) have similar compositions to chondrites, but show evidence of high temperature melting: hence they generally lack any chondrules and typically have a crystalline structure.

There are just over 140 members of the **Acapulcoite/Lodranite** group, comprising equal numbers of fine-grained Acapulcoites and coarser grained Lodranites. It is considered highly likely that the two sub-groups originated on the same asteroid, the former from the surface layers, the latter from deeper within the parent body. The type specimens were seen falling near Acapulco, Mexico in 1976 and Lodhran, Pakistan in 1868.

The thirty one currently recognised **Winonaites** are named after a 24kg meteorite that was reputedly discovered wrapped in a feather cloth in a Sinagua burial cyst in Arizona, USA. The group are thought to derive from the surface layers of the same planetissimal that was the origin of 1AB and 111CD iron meteorites: however, a few winonaites have been found to contain indistinct chondrules and might have formed on a different body.

The discovery that it is possible to assign a probable body of origin to some meteorites began with study of the fascinating HED clan, comprising three types of achondrite (**Howardites, Eucrites and Diogenites**) These originated in the surface regolith or crust of a fully differentiated body and have chemistries and lithologies similar to terrestrial igneous rocks. It was discovered that their relative reflectance spectra (see glossary) closely match that of the asteroid 4-Vesta, which displays a massive crater that is now generally considered to be the site of the impact that projected the HEDs out into the Solar System. This event took place around a billion years ago, creating the V-type asteroids of the Vesta Group and scattering smaller chunks of debris further afield. It seems likely that many of the HEDs that occasionally strike the Earth originated on a handful of small V-type asteroids that, through subsequent collisions, have entered into near-Earth orbits.

Eucrites are essentially basaltic rocks, most of which originated in the crust of 4-Vesta. There are three main groups: non-cumulate, polymict and cumulate, the first two of which originated at or near the surface, while cumulate eucrites are considered to have formed deeper in the crust within magma chambers. In total there are currently around twelve hundred known eucrites, making them the most abundant (and least expensive!) of the HEDs.

Diogenites are likely to have formed deep within the crust of 4-Vesta and contain crystals of orthopyroxene, plagioclase and olivine that are larger than those found in eucrites (suggesting slower cooling and growth) There are currently around four hundred recorded diogenites, arguably the best-known being the attractive grey-green Tatahouine meteorite that fell in Tunisia in 1931.

Howardites are brecciated rocks that formed on 4-Vesta when eucrite and diogenite ejecta in the regolith were compressed by the weight of overlying layers from subsequent impacts. Around three hundred and fifty howardites have so far been recorded, making them the scarcest and most difficult of the HEDs to acquire.

Eucrite (NWA 2949)   Diogenite (NWA 7831) Howardite (NWA 2696)

Plate 4: Slices of HED meteorites from 4-Vesta

Ureilites are carbon-rich achondrites: the presence of tiny diamonds in the group makes them both intriguing and difficult to cut! It seems likely that ureilites were liberated by a collision between two asteroids, one of which had a carbonaceous composition. There are around five hundred known ureilites, which are divided into coarse-grained olivitic monomicts and polymicts made up of unrelated clasts.

Brachinites are a scarce group comprising just over forty known specimens that includes several that, given their relative oxygen isotope (ROI) concentrations, may possibly have originated on the primordial Earth! The type meteorite, Brachina, was found in Australia in 1974 and originally classified as a martian chassignite (see below!) However, further study led to it being elevated to the type of its own group, whose origin is currently unknown.

Aubrites generally resemble some lunar types, having greyish, brecciated interiors and dark brown fusion crusts. The group is named after the town of Aubres in France, where the type meteorite fell in 1836. Aubrites are rich in magnesium and low in iron, typically being composed of a breccia of enstatite, olivine, troilite and other igneous minerals. Several members of the 'E-type' asteroid group have been suggested as possible parent bodies. So far around eighty angrites have been recovered, including al Haggounia 001, a 'fossil' meteorite which is so abundant that ownership of an example of this group is affordable to every collector!

Angrites are strange, rare achondrites, with just twenty eight examples currently known. Named for the Brazillian Angra dos Reis fall of 1869, these curious meteorites are chiefly composed of augite, olivine, troilite and anorthite. Their extreme crystallisation age (4.55 billion years), chemistry, ROI and presence of vesicles has caused a number of researchers to suggest an origin on the planet Mercury for some members of the group, in particular, NWA 2999.

Lunaites are arguably the most sought-after additions to any meteorite collection: whether the conspiracy theorists are right or not, the two sub-groups of lunar meteorites have **_definitely_** arrived on the Earth from its satellite, the Moon! There are over three hundred and fifty named lunaites, but many of these are either paired or twice-named: it's possible that the actual number is closer to one hundred and fifty. The total mass of these is around 220kg, which compares favourably with the 380kg brought back by the Apollo missions at a cost of $23 billion!

The two divisions of lunaites are those from montane (highland) regions and those originating in the flood basalt maria basins (the Moon's 'seas').

All lunaites are of great scientific importance because of what they reveal about the creation of the Earth-Moon system: since all the Apollo missions touched down in relatively smooth maria regions, the montane meteorites are of particular interest to researchers (and, in any case, hardly any of the Apollo rocks have been made available for study by independent scientists).

Cosmic ray exposure dating has shown that meteorites from the Moon have been arriving on Earth for millennia: the 'youngest' lunaite so far known, Kalahari 008, left the Moon just a few hundred years ago, while others are millions of years older. Some lunaites must have spent a considerable period orbiting the Sun before encountering the Earth, while others must have fallen directly here after launch. Given the Moon's low escape velocity of just 2.38km/sec, all are assumed to have been projected from the lunar surface by crater-forming impacts. The fact that several multiple pairings of different petrological types have been identified suggests that the actual number of impacts may be considerably less than suggested by the number of named lunaites. Recent research suggests that the majority of lunaites may have been launched by comparatively small events that produced craters just a few hundred metres across. The reasoning behind this is the fact that it is a long time since there has been enough of the very large 'rogue' asteroids and comets in Earth-Moon crossing orbits that would be required to form huge craters such as Clavius or Bailly.

In theory, there should be a 50:50 split between lunaites from the hemisphere that permanently faces the Earth and that which does not. In reality, we know from various lunar orbiters and the Apollo missions that the 'back' side of the Moon has far less large maria basins and more montane regions, so iron-rich mare basalt meteorites or basaltic breccias are more likely to originate on the 'near' side, while iron-poor feldspathic types were launched from the 'far' side. (Pink Floyd notwithstanding, there is no such thing as a 'dark side' of the Moon!) It is often forgotten that the landing sites for the Apollo missions were chosen to be 'petrologically interesting' so that the rocks collected are not typical of the maria basins. On examining the material from Apollos 12, 14 and 15 it was discovered that the general region of the manned landings is close to the **Procellarum KREEP Terrane** (PKT) which has rocks with unusual lithologies containing atypical amounts of potassium, rare earth elements and phosphorus. (So that the conventional distinction between the chemistry of montane and mare-basin minerals is not strictly valid.) This fact has allowed several lunaites to be identified as candidates for an origin in the PKT, examples

Plate 5: Slices of two different lunar meteorites

being Dho 1442, SaU 169 and others with unusually high amounts of the element thorium. Other lunaites have been linked to specific craters: the anorthositic fragmental breccia Dhofar 490 is similar in lithology to material collected during the Apollo 16 mission: it is considered possible that Dho 490 (and two paired meteorites) originated in the Highland region around the crater Descartes.

As is generally known, unauthorised possession of lunar material brought back by the six Apollo missions is **totally** illegal, so buying a lunar meteorite is by and large the only way to add a piece of the Earth's binary twin to a collection. However, there are a few exceptions to the rule – if you have the funds!

Launched in September, 1970, Luna 16 was the first spacecraft to land on the Moon, collect samples of dust and rock, and return them to Earth. After collecting regolith samples from the Sea of Fertility, where it had touched down, Luna 16, was launched back into space 26 hours later. It returned to Earth, bringing back 101 grams of Moon rocks.

Amazingly, a tiny amount of this material became available to collectors, albeit at mind-numbing cost! In December 1993, the auction house Sotheby's sold a slide with three small lunar fragments from Luna 16 for $442,500.

Following the six successful American moon landings (1969 – 1972) the Nixon Administration gave a total of 270 samples of material brought back from the Moon to each US State and to 135 member-countries of the United Nations. Astonishingly, around 180 are unaccounted for and at least one was the subject of a private sale at a price of $5 million. This, the Honduran Apollo 17 Goodwill Moonrock, was the subject of a 'sting' operation to retrieve the missing Apollo samples. After lengthy legal deliberations, the rock was remounted and presented back to Honduras in 2004.

Despite all the incredible security that surrounds the Apollo rock samples, there's quite a list of samples that have gone missing or have been lost irretrievably. These include specimens formerly held by Delaware & New Jersey (USA) Brazil, Canada, Cyprus, Ireland, Malta, Nicaragua, Romania, Spain and Sweden!

Probably the most intriguing of these is the Irish Apollo 11 sample. This was housed in the country's main observatory, at Dunsink, North Dublin. In October, 1977, an unexplained fire destroyed the library at the observatory: during the investigation and clean-up that followed, the Moonrock was apparently consigned to the nearby Finglas rubbish dump, where, presumably, it remains. There is, of course, another explanation: that someone stole the rock before setting the observatory on fire to cover his tracks and hide the theft....

The most-frequently asked question (both in relation to the Apollo samples and lunaites) is: 'Are lunar minerals unique to the Moon?

Lunar Dust from the Hadley-Apennines Apollo 15

This Certificate of Authenticity certifies and affirms that this material is an authentic sample of lunar dust originating from the Hadley-Apennine lunar landing site.

This lunar dust material was discovered ingrained into the interior of a Beta Cloth Temporary Stowage Bag (Illustrated at the bottom of this certificate) that was flown on board the Apollo 15 Command Module "Endeavour". This bag was acquired from a leading auction house in the USA where it was publicly sold after having been released from the North American company inventory. Three of these stowage bags were flown - one for each astronaut, and this specific bag (R-1) was used by mission commander David Scott and was attached to the left girth shelf next to his couch. This Stowage Bag was used to store all manner of small items and dry refuse which has resulted in a contamination of the interior of the bag during the voyage home by the lunar material which was ever-present in the capsule and on the clothing and bodies of the astronauts themselves.

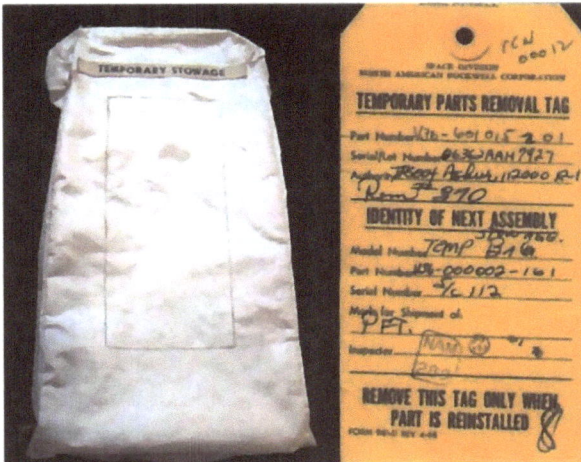

Plate 6: Lunar dust sample from Apollo 15

Sadly, the answer since 2011, is no! Examination of the Apollo moon rocks has revealed that most of the material is virtually identical to terrestrial anorthosites, basalts dunites and so on. Three minerals, however, were identified as being unknown on Earth: armalcolite, pyroxferroite and tranquillityite.

29

Since 1972, however, each of these minerals has been discovered on Earth, the last being tranquillityite, which was found in six localities in the Pilbara region of Western Australia.

This, naturally, has provided ammunition for the 'Apollo Conspiracy' debate: however, chemical composition aside, there are other differences, the lack of any hydrated minerals being the most significant.

As a matter of fact, one last potential source of authentic moon rock – or, to be exact – lunar regolith **does** exist! On his return to Earth, one of the two Apollo 15 Moonwalkers, Dave Scott, apparently found dust from the lunar surface inside his personal preference kit (PPK) This was a beta cloth bag containing a few sentimental items that each moonwalker was allowed to take with him on the descent to the Moon. Some astronauts took photos of their families, others rings and some even Masonic regalia! Scott had (supposedly accidentally!) transferred dust from his EVA gloves to the bag. A German collectibles dealer obtained some of this material on pieces of adhesive tape and sold it from his website. I myself had an example and can confirm it exhibited the classic 'orange marbles' look of the lunar regolith samples from Apollo 15.

Martian meteorites are about as exotic as a collector could wish for! As recently as when my wife and I studied for our astronomy degrees it was still not widely accepted that meteorites could actually reach Earth from the planet Mars: indeed, I once offered Sir Patrick Moore decent-sized samples of a lunaite and an NWA 4766, a martian shergottite as 85th birthday presents. To my surprise, he refused the shergottite, saying: "I don't believe for one minute that it's from Mars!"

Once the 'floodgates of acceptance' were opened by the discovery of increasing numbers of lunaites and other planetary types, it was inevitable that meteoriticists should start searching for meteorites from Mars. If impacts on the Moon could launch rocks into space, why, it was argued, shouldn't similar events have occurred on Mars? Since the escape velocity of this small world is only around 5km/sec and its atmosphere is comparatively thin, not only was it considered possible that impact events could generate Martian meteorites, but also past eruptions of some of the planet's huge volcanoes!

As is often the case in science, Martian meteorites had been gathering dust, unrecognised, in museums and laboratories for decades, in the form of the SNC ('snick') achondrites. S, N and C are the initial letters of three historical falls: Shergotty (India, 1865) Nakhla (Egypt, 1911) and Chassigny (France, 1815) By the early 1980s it was becoming obvious to a number of researchers that these meteorites (and half a dozen or so others found to have similar compositions) were sufficiently 'different' to other meteorites to be strong candidates for a Martian origin. They were all comparatively young, contained aqueous weathering products, shared the same oxygen-isotope compositions, and contained **mafic** and **ultramafic** minerals (see glossary) that suggested an origin in the igneous regions of a large planetary body.

Data collected by the various 'robot' explorers that have roamed across Mars in the intervening forty years have confirmed that the hundred and ten or so meteorites of the SNC and two other groups are indisputably Martian in origin: it is even possible to suggest a location for the impacts that generated many of them.

Arguably the most famous (notorious?) Martian meteorite is ALH 84001, found in the famous Allan Hills, Antarctica in 1984. This near-two kilo stone was found on examination to have left Mars and travelled around the Solar System for sixteen million years, arriving on Earth thirteen thousand years ago. It has an entirely different lithology to the SNCs and has been assigned its own grouplet, the **orthopyroxenites**. ALH 84001 is one of the most ancient Martian meteorites yet discovered, its component minerals being over four billion years old. At the time they were formed, Mars had sizable bodies of free water as well as sub-surface aquifers and a denser atmosphere rich in carbon dioxide: this is reflected in the presence of carbonates in the meteorite.

As if all this wasn't exciting enough, a group of scientists who had been studying the intriguing rock announced in 1996 their discovery of microfossils, which they considered to be bacteria. This proved to be somewhat precipitate and I'd bet they wished they'd held back a little longer when the then US President, Bill Clinton, hijacked their discovery for political ends. Further research eventually tended to disprove an organic origin for the 'bacteria', although it is the case that the discovery of magnetite crystals indistinguishable from those associated with those deposited by terrestrial bacteria has continued the debate.

In 2011 a completely new type of Martian meteorite was discovered, allegedly in Morocco. NWA 7034 proved to be an ancient basaltic breccia and was found to have the largest amount of water yet found in any meteorite, in the form of hydrated minerals and clasts of quenched magma. So different is NWA 7034 that an entirely new grouplet was established to accommodate it! It is considered highly likely that the igneous material from which the meteorite originated was erupted into a Martian sea just over two billion years ago, making NWA 7034 the second oldest known Martian meteorite. Its rarity and colour have led it to being nicknamed 'Black Beauty' by many collectors!

We turn now to the stony irons, a family that contains some of the most beautiful and scientifically important meteorites.

As a professional dealer in space rocks, it is perhaps unsurprising that I find all meteorites equally fascinating and, in their own way, aesthetically appealing. But I have to admit that **Pallasites**, with their beautiful structure of fragments of olivine (or peridot, as gemologists refer to it) suspended in a nickel-iron matrix, are probably the most visually exciting, particularly to the non-specialist.

In addition to their undoubted beauty and rarity, the four groups of Pallasite offer an intriguing glimpse into the interior of a planet that make them among the most scientifically important of all meteorite types.

Main group Pallasites (PPM) are named for the incredible German polymath, Simon Peter Pallas, who, having been shown an unusual 650kg mineral specimen while visiting the Russian city of Krasnoyarsk in 1772, was instrumental in its recognition as a new class of meteorite. The city of Pallasovka in the Volgograd region was also named in his honour (as were dozens of plant and animal species) In one of those delightful coincidences that occasionally occur, a new pallasite (now named Pallasovka) was discovered there in 1990.

There is still some debate about the actual origin of Pallasites: although some meteoriticists hold that their stony-iron structure resulted from collisions between nickel-iron asteroidal cores and chunks of mantle material, it has also been suggested that they originated in the core-mantle boundary layer of differentiated planets that were shattered during collisions.

Examination of the hundred and fifteen or so known pallasites is strongly suggestive of the probability that they were released into space from a number of *different* collision events: there is considerable variation in their chemistry, structure and ages – from over four billion to just a few hundred million years. (This, of course, reflects the time of the impacts that released them into space rather than of their formation during the differentiation process of the body of origin.)

The purity of the olivine crystals in some pallasites is such that those from the 45 kg Marjalahti meteorite were adopted as the official standard for peridot (the gemological name for crystalline olivine) The high prices commanded by many pallasites reflects the fact that meteorite collectors are competing for them with jewellers and gemologists!

Undoubtedly among the most beautiful (and stable!) of all Pallasites are Imilac, from Argentina and Esquel, from the driest desert on Earth, the Atacama in Chile. Unfortunately, both are prohibitively expensive, at upwards from £50/gram! A decent slice of either would command a three-figure sum...

Fortunately there *are* stable meteorites that are reasonably priced: these include Pallasovka, which has large brown-orange crystals and Jepara, discovered during the building of a furniture warehouse in Indonesia in 2008. Jepara is an ancient fall in which the nickel-iron matrix has gradually transformed into magnetite, schreibersite and nickel sulphide. Since no further oxidation is possible, Jepara is very stable. Very thin slices display a beautiful structure of greys, silvers and browns, with pale yellow olivines. All the slices I have imported from the owner of the main mass have been treated with **opticon** and have been totally stable.

Other 'fossil' Pallasites turn up from time to time: the best-known is Huckitta, found in Northern Territory, Australia in 1924. Here the oxidation process is virtually complete: the nickel iron has oxidised to silver-grey

Plate 7: Jepara pallasite

magnetite and haematite, while most of the olivine has usually degraded into dark grey crystals. Other similar Pallasites appear on the market from time to time, notably the recent North West African examples NWA 4482 and NWA 6576. Although lacking the stunning impact of those containing translucent olivine crystals, these ancient falls are still fascinating and attractive in their own way: and, of course, they are utterly stable!

At the time of writing, slices of the generally stable Russian Seymchan pallasite are still widely available: the translucent olivine fragments in a good example of this 1967 find are frequently of three different colours: green yellow and reddish orange. When I hold a slice up to a bright light I am often reminded of traffic lights or wine gums!

The internal structure of the original Seymchan pallasite is taken by many meteoriticists (including me!) as evidence of a core-mantle boundary origin: some regions are complete devoid of olivines, while in others the crystals are jammed together with hardly any metal. Still others seem to represent transitional regions between the nickel-iron core and the olivine mantle of the disrupted parent body.

Plate 8: Seymchan pallasite slices showing transition

A recent addition to the inventory of main group pallasites has been the Sericho (occasionally Habaswein) Kenyan find. Several tonnes of this Pallasovka-like pallasite were discovered in 2016 and rapidly become available at dramatically affordable prices: as low as £1/g. At the time of writing, it remains to be seen just how stable Sericho will prove to be, given its origin and the fact that much of the material is in small, whole specimens. (Larger meteorites tend to be less weathered internally than small ones: as we'll see shortly, that is particularly the case with the Campo del Cielo iron.)

The **Eagle Station** grouplet (PES) consists of just five members, named for the type specimen found in Kentucky in 1880. The grouplet all have characteristic oxygen isotope concentrations and contain olivine crystals that are unusually rich in calcium and iron.

There are presently seven known **Ungrouped Pallasites** (P-ung) which do not 'fit' into the classifications above, either because of metal or oxygen isotope compositions. Some meteoriticists suggest that two of these are sufficiently distinct to merit their own subgroup:

The **Pyroxene Pallasite** grouplet (PPX) currently has just two members, both of which contain pyroxene as an accessory mineral and possess distinctive OICs. However, there are significant differences between the two (Vermillion, found in the US in 1991) and Yamoto 8451 (Antarctica, 1984) to suggest that the formation of a separate grouplet is not justified.

The **Mesosiderites** make up the second division of the stony iron meteorites. Like the pallasites, their composition is more or less half silicate material and half nickel-iron: however, their structure and likely formation is completely different.

Whereas pallasites consist of a metal matrix in which silicate minerals are suspended, the opposite is the case with mesosiderites. These are basically stony meteorites with variable amounts of metal inclusions. The stony matrix is eucritic in nature, indicating an origin on the surface of a planetary body: this is taken as strong evidence that mesosiderites were formed by collisions between differentiated asteroids and the metallic cores of disrupted bodies. (Interesting thought: it could well be that a future find will be composed of eucritic and pallasitic material. How impressive would *that* be?)

Possibly the best-known examples of this group are **Vaca Muerta** from Chile and **Bondoc** from the Phillippines.

Plate 9: Mesosiderites - Bondoc and Vaca Muerta

We have arrived at the final major group of meteorites: the **nickel-irons**. (For convenience, 'irons' from now on!) When most people hear the word 'meteorite', the first image that forms in their mind's eye is of a big lump of metal whose surface is covered with shallow depressions like thumb-prints in wet clay. Whenever meteorite impacts or extinction events occur in TV programmes or in films, an iron meteorite is generally what is pictured: whenever I display a selection of meteorites at a lecture or mineral fair, it is the large irons that always attract the most interest. With their dark, lustrous surface and high density, irons conform to our ideas about what something that came from space should look like!

The origin of iron meteorites remains the subject of ongoing discussion and investigation. It is considered that the majority are fragments of the cores of small, differentiated planets that were shattered by massive collisions around 4.2 billion years ago: these were most likely members of the M-asteroid group (based on surface spectroscopy) However, there is a suggestion that nickel-iron minerals may have also condensed directly in widely dispersed regions of the solar nebula: only the 111AB group formed in the main asteroid belt, while groups IAB and IIAB formed nearer the Sun, and group IVA may have an origin outside the orbit of Jupiter. A further group, the 11E irons, seem to have their origin in the crust of the S-type asteroid, 6 Hebe. (The number of asteroids that have been linked to various iron meteorites suggests that there must have been a lot more of them earlier in the history of the solar system!) Such non-magmatic irons may have formed on the surface of these asteroids as 'pools' of nickel iron left by high-energy impacts.

There are three sub-groups of irons, which together total just under 1200 examples, around 8% of all known meteorites: because of their great density (7-8 g/cm^3) it is a surprising fact that irons make up 85% of the total mass! The main distinctions between the three types of iron are the size and shape of their crystal structure and the ratio between the amounts of nickel and iron in their composition (largely in the form of the two minerals kamacite and taenite) When polished and etched (see appendix), some irons (and pallasites) display attractive crystal structures: this is named after Count Alois von Beckh Widmanstätten, its discoverer in the early years of the nineteenth century. The size and nature of the Widmanstätten pattern where present is also used to classify irons. Virtually all other metallic elements occur in iron meteorites in tiny amounts, the next most abundant being cobalt. The presence of metals such as iridium that are extremely rare on Earth is often confirmation that a crater or mineral deposit is meteoric in origin.

Hexahedrites are almost pure kamacite, containing <5% nickel. The faces of hexahedrite crystals (revealed by super-cooling and shattering) display fine parallel striations known as Neumann Lines. These irons don't display a Widmanstätten pattern (WP).

Octahedrites contain a higher percentage of nickel – from 7% to 10%. These meteorites contain both kamacite and taenite and display a Widmanstätten pattern: the width of the WP bands varies from fine to medium to coarse to most coarse, the higher the nickel content, the narrower the bandwidth. Really coarse octahedrites (such as the well-known Campo del Cielo from Argentina) include very large crystals of both kamacite and taenite and display an attractive crystal structure when etched and Neumann lines on the faces of the kamacite crystals when shattered.

Plate 10: Widmanstätten pattern and Neumann Lines

Those of you who had chemistry sets when young (in the days when children were still allowed to be amateur scientists at home!) will recall that growing large crystals (such as sodium chloride or copper sulphate) requires patience and time: the longer you allow a crystal to grow in a concentrated solution, the larger it will become. This is also true of the crystals in iron meteorites: it is considered that the large crystals in Campo del Cielo must have required thousands of years to grow – evidence of the vast kinetic energies released by the impact that liberated them from a planetary core.

The final division, the **Ungrouped Irons**, consists of iron meteorites that do not fit into the main chemical classifications: many of these were formerly referred to as Ataxites. The group includes irons with the highest amounts of nickel in their composition: >18% and, with just one hundred and twenty four currently-known examples, are the least abundant of the irons. Interestingly two historically important and large irons – Hoba and Chinga – are both examples. A recent addition to the group was made in fascinating circumstances: in 2009 an Italian-Egyptian geophysical team investigated a possible impact crater near the Gebel Kamil (a small hill in the East Uweinat Desert, Egypt) that had been observed on 'Google Earth' by an Italian researcher. They discovered that the comparatively small forty five metre crater contained over sixteen hundred kilos of irons with masses ranging from 1g to 35 kg. These all display a beautiful brown outer layer that resembles shark or lizard skin.

In addition to the system above (based on crystal structure) iron meteorites are frequently classified by their chemical compositions: this is the system most generally used by meteoriticists and researchers.

Structural classification	Chemical classification	Mineral components
Om-Og	1AB	kamacite, taenite, silicates, carbides
Om-Og	1C	kamacite, taenite, silicates, carbides
Ogg-H	11AB	kamacite, taenite
Ogg	11C	kamacite, taenite
Of-Om	11D	kamacite, taenite
Off-Og	11E	kamacite, taenite, silicates
Plessitic/Oct/Ataxite	11F	kamacite, taenite
Om-Og	111AB	kamacite, taenite, phosphides, troilite
Off-D	111CD	kamacite, taenite, carbides
Og	111E	kamacite, taenite, carbides, graphite
Om-Og	111F	kamacite, taenite
Of	1VA	kamacite, taenite
D	1VB	kamacite, taenite
A11	Anomalous	kamacite, taenite, silicates, graphite

Plate 11: Classification of iron meteorites

And finally: **Ice from above!** A recent TV documentary (of the somewhat popularist variety!) sought to explain several recent – and very damaging – falls of ice. Arriving at high speed from cloudless skies, these have battered roofs, cars and aircraft. The conclusion of the program

44

Plate 12: Some different types of iron meteorite

Top: Sikhote Alin shrapnel   Agoudal/Imilchal   Gebel Kamil

Bottom: Campo del Cielo   Campo 'mega crystals'   Campo crystals

45

was that these were examples of 'mega-hailstones', poorly-understood phenomena, more usually called **megacryometeors** by the Scientific community.

Around 50 of these have been recorded so far this century, varying in mass from 0.5 kg to real giants such as a Brazilian example of over 50 kg: a specimen with a mass of 200kg was reportedly seen falling in Scotland in the nineteenth century!

Meteorologists on the documentary spent much of the program establishing a mechanism by which huge chunks of ice could form in the upper atmosphere, other than in the conventional nursery of the convection currents of a cumulo-nimbus cloud. As most people will be aware, the powerful updraughts inside such clouds (which are typically associated with thunderstorms) allow the formation of hailstones. These may gyrate inside the cloud, accumulating mass until they are too heavy to remain aloft: hailstones the size of golf balls are not that uncommon. However, they generally display a layered structure similar to that of an onion, while megacryometeors do not.

At the time of writing, no generally accepted mechanism for generating and supporting such large masses in the upper atmosphere has been forthcoming, although some of the theories put forward seem credible at first glance.

That chunks of ice occasionally fall from aircraft is undeniable and may be placed in two separate categories. The first (of which I have personal experience) is generated by the dumping of liquid waste from on-board lavatories. Some years ago a local radio station invited me to interview a lady who had been struck on the arm while hanging out her laundry. On arrival, I asked to see the object that had hit her: to my amazement, she opened her freezer and took out a polythene bag, inside which I saw a bluish lump of ice. The lady was less than thrilled when I told her she'd been storing lavatory waste from an aircraft among her fish fingers and frozen chips! It had an obvious smell of disinfectant, so I'm at a loss to know how the lady for one minute thought it was a meteorite...

Of course, icing on the fuselage and wings of aircraft still poses a threat to aviation safety, and flying through Supercooled Large Droplet (SLD) conditions can generate chunks of ice: these could fall to the ground and cause damage, but would not, I feel, be confused with genuine extraterrestrial ice meteors that should show signs of flowlines and ablation.

So then: is it possible that ice meteors could reach the Earth *from space* and pass through the atmosphere to the ground?

An online search will quickly discover a good number of learned publications that appear to answer this question with a resounding 'no'. The majority maintain that a vast initial mass of tens of thousands of tonnes would be required for a football-sized chunk to reach ground level. But is this necessarily the case? The assumption is that frictional ablation would melt away most of the object's mass, but this ignores certain factors:

1) Objects entering the atmosphere from deep space may hit the Earth head-on, at a combined velocity of 220,000kph or more. But equally, they may 'creep up' on the Earth from behind, with a closing velocity of just a few thousand kph, reducing frictional heating by a huge amount.

2) Our putative ice meteor would be at a temperature just a few degrees above absolute zero (-273 degrees C) Whilst the outer layers would indeed become extremely hot, they would slough off like the heat shield of a re-entering Apollo spacecraft, taking heat energy with them.

Moreover, like the tiles on the five Shuttle Orbiters, ice is a pretty poor conductor of heat, so that the interior might be expected to survive better than, say, a piece of rock or iron.

Assuming, then, that some of the ice that falls from the sky may indeed originate in space, two questions immediately occur:

Where would ice meteorites originate?

The Solar System is full of water-ice: billions of tons make up most of the mass of each of the trillions of comets in the Oort Cloud and Kuiper Belt. Additionally, the Asteroid Main Belt must hold thousands of captured objects from these remote regions: these, we know, are occasionally deflected into the inner Solar System. One of my previous books explored the possibility that the majority of large craters on the inner planets and their satellites are the result of **cometary** (rather than asteroidal) impacts: certainly, recent research has shown that to be the case with the Gilf-el-Kebir in Egypt.

How could we prove an extraterrestrial origin?

The relative concentrations of two isotopes of oxygen (^{17}O, and ^{18}O) is used to assign an origin to planetary and asteroidal meteorites. Cometary water *should* display relative oxygen isotope concentrations different to that of terrestrial water. At the time of writing, just a few megacryometeors have been tested: these have all had terrestrial ROICs. But the samples tested were just a tiny fraction of the numbers that fall on the Earth: it is unscientific to discount the possibility of ice meteorites on such a small sample.

Should a sample be found to have an exotic ROIC, it could then be examined for evidence of presolar grains, interplanetary dust and regalith fragments it had picked up during its wanderings in space.

In conclusion, I'd suggest there is a high probability that chunks of cometary ice do reach the Earth's surface from time to time: of course, the only ones that might be capable of collection and analysis would be those that fell on mountain tops or in the Polar Regions.

Starting and building a collection

IF I had written this book forty years ago, this chapter would have been much shorter: basically my advice would have been: 'Don't even try!' Back in the nineteen seventies there was little public access to the internet, no online auctions or payment systems and hardly anyone looking for meteorites in the World's remote regions.

By the nineteen eighties, European fossil collectors and dealers had discovered that the north-western Sahara was particularly rich in fossils from every geological period. From exquisite trilobites to fish to huge therapod dinosaurs, the region to the south of Morocco in particular produced a rush of new species, often represented by hundreds of examples. Most of these were found by tribal peoples such as the Tuaregs and Berbers, but these were soon joined by refugees from various conflicts in politically turbulent countries such as Algeria, Mauritania, Niger, Mali, Sudan and Libya. Dealers from Morocco in particular would regular visit refugee camps and purchase fossils in vast numbers for very little money: I recall how the price of Mosasaur teeth and jawbones plummeted at this time. The dealers soon discovered that their sources were also finding meteorites – lots of them! Against the desert sand and white limestone plains, incongruous black rocks were easy to spot and the dealers were quick to realise the potential rewards to be made. The 'Saharan Gold Rush' was underway!

Between 1990 and the early years of the twenty-first century, literally hundreds of thousands of putative meteorites were exported from North Africa. Not wishing to lose their new-found incomes, the discoverers were often loath to reveal the locations of their finds. Newly-established US and European dealerships (myself in-cluded!) were not too worried at the time, the rationale being:

- Without the demand we were generating, most of the new material we were importing would have remained unknown until it eroded to dust. Fresh falls / finds are of far greater interest than old, weathered ones.

- The money being sent by the west into villages and refugee camps ameliorated many of the problems of their poor inhabitants.

- Where a meteorite is **found** is of little significance: only the trajectories of observed **falls** can occasionally be used to derive an origin in a specific region of the Solar System.

Despite this, I recall being involved in a heated debate at the end of a lecture I'd just given to the AGM of a famous astronomical society. I was called a 'meteorite prostitute' by a very well-known meteoriticist, and was accused of being involved in meteorites purely for financial reward. I replied with the points above, adding that I'd given talks to countless school and college pupils (offering them the then almost unheard of opportunity to handle meteorites), donated sets to all kinds of institutions and written many articles that I hoped had increased public awareness and appreciation. I finished by reminding my critic that most of us – including University Dons – do what we do for money: it's called earning a living!

The international trade in meteorites grew at an incredible rate and it is absolutely true that several dealers in the USA very rapidly became millionaires. This wasn't so much the case in the United Kingdom: for many years there were only two of us attempting to make a full-time living from meteorite dealing and neither of us could be accused of being wealthy!. In Germany and France the situation was closer to that in the US and there are still some massive events on the continent where hundreds of meteorites change hands for thousands of Euros. (I'm not sure what this tells us about the British and the general interest in Science!)

Although the appearance in the 1990s of numerous websites offering meteorites for sale (including Spacerocks UK!) accelerated the rate of growth of popular meteoritics, it was the online auction phenomenon that dramatically increased both demand and supply. Yahoo!, Bidville and Auctionweb (later eBay) almost overnight became the default source of anything to do with space collecting.

As it happens, although I occasionally buy meteorites and tektites on an online auction site, I have never sold any on one. There are a couple of reasons for this:

- I had an existing clientele, online and at various public events (for a time fifty a year!) who bought most of the material I could obtain: I didn't want the pressure of having to source bulk stocks of meteorites in anticipation of increased seasonal demand.

- I preferred to remain a smaller-scale operation, so that I'd never have to worry about trying to sell the half a tonne of unclassified NWA chondrites gathering dust in the garage, nor about how to cover my on-costs and liabilities if and or when the bubble burst. (At collectors' fairs, I met a fair number of people who over-invested in 'beanie' toys in the 1990s, only to lose fortunes when the public moved on to the 'next big thing'!)

- I didn't want to become involved in the doubts cast about the authenticity and provenance of some of the items offered for sale in online auctions.

- I decided to set my **own** prices, based on agreed, fair payments to North African and other collectors, rather than become involved in price wars online. This has turned out to be a good thing, since the meteorites I sell are generally very competitively priced.

The object of this chapter is, you'll recall, to help people start or add to new or existing collections. It would be wrong to pretend that eBay isn't a wonderful resource for buying meteorites: many of the World's top dealers are regular sellers with online stores. What's more, there are real bargains to be had at times! Sometimes a vendor misspells an item or lists it in the wrong category: the item's potential buyers never come across it, so that it sells at a low reserve – or even less if no reserve has been set! On occasions a dealer forgets to stipulate where an item will be available for purchase and it ends up on an eBay site where there are few customers. On one memorable occasion I used the 'buy now' function to close the deal on a large Canyon Diablo iron that the seller had mistakenly listed at a fraction of its true value by putting the decimal point in the wrong place! (Under most sites' rules, once a deal is confirmed, that's that!)

To sum up this section, the watch-word is **caveat emptor** – buyer beware. Here are a few things to bear in mind:

- There are **always** plenty of fake or misrepresented items on eBay: in particular, be cautious about tektites, Moldavites, Libyan Desert Glass and rare achondrites. Never buy anything lacking a decent provenance. (Moldavites come from the Czech Republic, not China!)

- Watch out for low 'buy now' prices with hidden and outrageous handling and / or postal charges. This is in theory outlawed, but is nevertheless becoming increasingly common.

- Remember that anything you buy from outside the EU (and possibly from inside it after 'Brexit') will attract import and customs fees. These will greatly add to the price for which you think you've bought an item.

- When buying from online auctions, try to use well-known dealers: look them up on Google to see how long they've been active and if there's anything dubious in their past history.

- Look for the IMCA logo. This is **not** a guarantee of an item's authenticity, but membership of the International Meteorite Collectors Club is still pretty exclusive and members do look out for fakes, scams and stolen items offered for resale.

Of course, there are sources of meteorites other than eBay!

Plate 13: IMCA logo

Since we first started Spacerocks UK, my wife Linda and I have sold our meteorites at public events all over the country, the majority of these being rock, mineral and gem shows. At the peak of their popularity we attended up to fifty such shows a year, from Newcastle to Brighton and as far west as Exeter. Initially there was a long waiting list to become a trader at these events, many of which were held in large venues such as race courses and county showgrounds. When we first became involved, there were rigid regulations about what could and couldn't be displayed which had to be followed to the letter. The majority of the stands sold quality mineral specimens, fossils and raw or faceted gemstones, with occasional polished crystal and stoneware items.

Over the years attendances began to decline (possibly because of diminishing interest or as a result of the economic recession) and many of the events either closed down or changed ownership. Almost overnight the organisers of the remainder began to focus on increasing footfall and improving profitability. Table costs were raised to the extent that many of the smaller dealers could no longer afford to attend: they were frequently replaced by stalls selling esoterica such as wands, crystal balls and 'healing stones'. Sad to say, these once fruitful sources of high quality specimens (including meteorites) transformed into events that no longer attracted serious collectors. At the last one we attended, over a third of the stalls were selling beads and bead-making accessories and many more displayed polished crystal skulls, angels and chakra cleansing kits!

There are still a few authentic 'old school' mineral shows, some of which we attend with our meteorites: at the time of writing these include events in Collier Row, Oxford, Harrogate, Haywards Heath and Bakewell. For Linda and me, though, the thought of driving hundreds of miles on today's crowded motorways just to break even is no longer as appealing as it once was, so we tend to be more exacting these days!

The events we *do* still attend are the numerous astronomical shows and conventions that take place throughout the year. These range from local and national society AGMs to open days at universities and colleges: recently we have shown our meteorites in Chelmsford, Colchester, Coventry, Norwich and Cambridge. There are a couple of really high-profile two-day Astronomy

Plate 14: International Astronomy Show

Shows at which world-renowned authorities lecture on astronomical topics and trade stands sell a full range of telescopes, accessories, home observatories and books. These are usually staged in the Midlands or London and are extremely well-attended.

In some rural areas of the UK 'Star Parties' are held several times a year, when amateur astronomers set up their telescopes for a week of camping, lectures and communal observation. A few traders usually attend the final weekend of these events.

Of course, there are potential sources of meteorites other than astronomy and mineral shows. The first 'space rock' *I* ever owned was a small tektite, which, as a boy of ten, I bought in a cluttered old junk shop in Ilfracombe. It seems to me that there were a lot more of these back in the fifties and sixties, and some of my earliest memories of family holidays are of rummaging around in dusty boxes and cabinets: I still have the tektite and several other specimens I bought as a child!

If you manage to find an authentic curio shop like this (as opposed to one filled with 'shabby chic' and leather suitcases!), it's definitely worth a look. Some years ago I purchased a collection of meteorites that must have been assembled back in the early eighties: it included some real rarities including Allende, Miles, Gibeon and the largest slices of Imilac and Esquel I've ever owned! They had been sold to the shop by the widow of the collector at just a fraction of what I would have happily paid her… Sadly, we all have to pass on one day, and there's no guarantee that our heirs will have the time, knowledge or inclination to sell on our collections at market prices.

When you're on holiday at the seaside, you might come across rock and mineral shops, particularly in well-known fossil or mineral-rich regions such as the Peak District or the Jurassic Coast of Dorset. In the same way that rock and mineral shows increasingly offer little of interest to serious collectors, many of these now seem to concentrate on Harry Potter and crystal therapy-themed paraphernalia. However, there are still plenty worth browsing through. For example, there are a couple of cracking little shops in the back streets of Cromer (Norfolk) and Lyme Regis (Dorset) that frequently offer decent meteorite samples for sale, often at reasonable prices – so much so that I once bought several kilos of Nantan irons and a bag of moldavite from one of them!

Finally, it's definitely the case that meteorite and mineral collecting are 'bigger business' in both the USA and in several European countries: this is reflected by the scale, scope and attendance of a number of overseas events:

- **The Tucson Show**
  This annual event (spreading across January and February) is actually a cluster of over forty shows, situated in exhibition halls, hotel foyers, roadside stalls and even hotel rooms! Dealers, exhibitors and customers travel from all over the world and much of what is offered for sale on eBay originated at Tucson!

- **Ensisheim**

  This event takes place each summer in the small Alsace town near which a large meteorite fell in 1492. Many dealers and collectors regularly attend, more as a social event than anything else. However: trading and exhibitions do take place.

- **Munich**

  The Munich Show is probably the best place to examine and purchase meteorites. Taking place each October, the event attracts specialist dealers from around the globe, including a good number from Morocco. In the past, real bargains could be found in the 'unsorted' bins of wholesale dealers: even now, Munich is the venue for newly-discovered or identified material.

- **Paris**

  An annual December event since 1971, the Paris Mineral Show takes place at the 'Espace Charenton' in south-eastern Paris, attracting over 200 dealers, many of whom sell meteorites.

Curating and displaying a collection

HAVING spent time and money assembling a collection of meteorites, it obviously makes sense to protect your investment with a little TLC!

I recall some years ago selling a beautiful, large slice of the Brahin pallasite to a new collector in London. I spent some time explaining how he should care for it and gave him an A4 sheet telling him exactly how to do so. About a year later he phoned up to complain that his item had disintegrated into a pile of rusty metal and olivine crystals! Naturally I was very concerned and asked him how and where he had displayed it. He replied that he had put it on a copper plate stand on the window sill of his bathroom! It's hard to imagine a worse way to treat a piece of polished nickel-iron, apart from putting it on the rockery! He confessed that he had never got round to reading the fact sheet! It's worth bearing in mind that on Earth the fate of all polished iron is to go rusty!

Rusting is actually a quite complex multi-stage reaction during which iron atoms first react with water to form iron

Plate 15: Before and after!

hydroxides which then combine with oxygen to become iron oxides – rust. The thing to notice is that iron can only become rust in the presence of water: iron meteorites discovered on Mars by the Curiosity rover show no sign of rusting whatsoever.

When an iron meteorite lands on Earth it is most likely to land somewhere wet, not only because 70% of the surface is covered by seas, lakes and rivers, but also because most of the dry land is either temperate or tropical. Any meteorite landing in these climate zones that is not recovered straight away will immediately begin to deteriorate. If the meteorite is an iron or pallasite, once a thickish external rust layer has formed this will tend to protect the inside of the meteorite to some extent. However, irons with large crystals such as Campo del Cielo tend to continue rusting as water percolates down through the wide crystal boundaries, while cutting and polishing any meteorite will expose the inner structure to water vapour in the atmosphere.

Ideally, then, any iron meteorite or pallasite that you add to your collection should originate somewhere totally arid, such as a desert or polar region. This is, of course, not often going to be possible, which is why meteorite collectors need to find a way to halt – or, if possible, reverse – the rusting process.

The first, most obvious option to consider is housing your meteorites in airtight acrylic boxes with a sachet of silica gel or similar desiccant. In the short term, this works quite well, but there are a couple of obvious drawbacks:

- Sealing the box prevents you handling the meteorite for an occasional close examination.

- Once the silica gel has become fully hydrated, not only does it no longer absorb water, it actually gives it off! If you don't regularly check the sachets (which change colour as they absorb water) and replace them as soon as they are saturated, your sample might quickly rust away.

Many meteorite importers prefer to stabilise newly-arrived irons and pallasites and then seal their surfaces to prevent (or, to be honest, slow down) the rusting process. This is generally achieved in one of two ways. Firstly, the meteorite can be soaked in isopropyl alcohol (often sold as rubbing alcohol or surgical spirit by chemists) for a few days: this will dehydrate both its surface and

Plate 16: desiccant crystals

any small cracks. Next, heat the meteorite to 70°C to evaporate the alcohol, before sealing it to prevent water condensing on it again. At one time, American dealers seemed particularly fond of using cobblers' wax to seal the outer surface: the blackish Campos you see on eBay have been treated in this way. Personally I prefer to use a mixture of two lubricants: Duck Oil and Wax Oil. After a couple of days in this the meteorite can be dried off and wiped with a dry cloth, producing a more natural slight sheen. Treated in this way, even small Campos (which seem to be the most prone to rusting) usually become pretty stable.

All metal meteorites benefit from the occasional application of a surface film of light oil such as gun or sewing machine oil or '3-in-1 ' A word of warning, however: don't be tempted to spray your prized specimen with WD-40.

Plate17: Campo del Cielo irons before and after cleaning

As excellent a lubricant as this is, it is also hydrophilic, and actually draws water in from the atmosphere. (This is, of course, why it is sprayed on car electrics as a 'damp start'!)

The items most prone to deterioration are polished and etched slices, not the least because both processes require the use of water! Needless to say, freshly prepare samples must be very carefully dehydrated with isopropanol and then dried in an oven. Often they are then sealed with an acrylic coat: some dealers use clear nail varnish, others a proprietary sealant called **opticon**. Both work very well, but do have the drawback of sealing any remaining moisture inside with no way for it to escape!

As an experiment, I have recently begun to treat newly-polished pallasite or iron slices with a proprietary brand of jeweller's metal burnisher and wax polish: results so far look optimistic: there is an extra advantage in that the specimens acquire a pleasant smell!

Plate 18: Jeweller's burnisher and polish

Because of their often high iron content, many stony meteorites need a little care, too. The nickel-iron in L chondrites, (such as the Ghubara L5, for example) is concentrated in small flecks which look really attractive in a mirror-polished slice. Unfortunately, if untreated, these can rapidly turn into oozing, rusty spots that stain their display boxes.

All the smaller slices and individuals that we sell are

supplied in padded acrylic boxes: sachets of silica gel (and data labels) can be inserted underneath the foam padding if desired. The boxes themselves are pretty much airtight, but I sometimes put iron samples in small 'ziplock' bags, both to further inhibit deterioration and also to protect the white padding. Some sizes of these boxes are becoming harder to find these days: at the moment I import mine from Schmelzer, a German wholesaler. Anything bigger than a conker generally has to take its chances!

Some of our customers keep their collections in purpose-built hermetically sealed display cases, with either an electrical dehumidifier or one of those boxes of crystals that are intended for damp rooms or wardrobes. These have the advantage of being very cheap, but must be regularly checked and replaced when the crystals have become fully saturated. Other clients ask me to frame their prize samples in hardwood frames that are made by a local cabinet maker. With an informative printed backing card, these not only look terrific when hung on a study wall, but also provide additional protection. Still others house their collections in traditional multi-drawer cabinets. Whatever your personal preference, it's always best to house your meteorites in the driest room in the house – and don't forget to check them regularly!

Unless you intend to put together a representative 'type set' containing one of each group, a collection of meteorites from a particular country or continent (or all of them!) or specimens from the decade or year of your birth (all of which have been achieved by some of my clients!) then it may be easiest in the long run to acquire only meteorites that are known to be 100% stable: there are enough of these that my wife, Linda is able to maintain a large inventory of meteorite jewellery containing examples of all the major groups.

Meteorites have proven themselves to have been a wonderful investment in recent years, but it's absolutely essential that you keep a record of the provenance of each item in your collection: otherwise they might just be seen by your executors as a bunch of pretty stones that may end up on the rockery! Ours are always supplied with an A4 or A5 factsheet (and certificate of authenticity) but if yours don't have these, it's worth producing your own catalogue, either hard copy in a folder or on your computer's hard drive.

Plate 19: Our factsheets and box labels

Where can I see some meteorites?

AT my increasingly-occasional public lectures and sales events (I'm getting old!) I am frequently told by professional and amateur astronomers alike that my meteorite display is the largest they've ever seen (and had the opportunity to touch).

Perhaps surprisingly, in the UK there are not many large collections of meteorites on public display. The Natural History Museum in South Kensington, London has an impressive two thousand or so specimens, but, at the time of writing, the vast majority are kept 'behind the scenes': academics and serious researchers are occasionally granted access, but the public only get to see a few representative examples. (Which, it has to be admitted, are all pretty spectacular!) I was recently informed that this situation will change when the Museum finds enough money to display the collection: considering the amount recently spent on hanging a whale's skeleton in the entrance hall, this, I feel, says something about the priorities of the management.

Several provincial museums have a few meteorites on

Plate 20: Spacerocks display

permanent display, some of which are well-worth making an effort to visit:

- The Sedgwick Museum, Cambridge has a few nice specimens on show, as well as some interesting material from the Apollo program.

- The National Space Centre, Leicester has several meteorites, among which is a large Campo del Cielo that can be handled, as well as occasional other specimens of rarer types such as Martians and Lunars.

- There is quite a large collection at the University Museum of Natural History, Oxford, of which a number of interesting examples are on display. These include several irons and pallasites, as well as an 8.5kg sample (the second largest in existence) of the Limerick H5 stone that fell in 1813.

- The Manchester Museum has good examples of some well-known and important meteorites, including Bondoc, Appley Bridge, Millbillillie, Allende and Odessa.

- The Yorkshire Museum has a small collection that is not always on display: among these is the beautiful, oriented 1600g Middlesbrough L6 chondrite that fell in 1881. Check that it's on show to the public before visiting: if it is, it's well-worth the effort!

- The Hunterian Museum in Glasgow has a collection of seventy or so meteorites, among which is the largest remaining chunk (150g) of the High Possil L6 chondrite that fell in Glasgow in 1804.

- Just a few of the large collection of meteorites held by the National Museum of Scotland, Edinburgh are currently on display: these include slices of Esquel and Muonionalusta and examples of Nakhla and the Chelyabinsk fall.

- Perth Museum holds a one kilo fragment of the Strathmore L6 chondrite that fell in 1917. This is an example of what the Americans call a 'hammer stone': pieces of the meteorite smashed through the roof of South Lodge, Keithick.

- The National Museum of Wales, Cardiff has a small collection of meteorites, including a fragment of the Beddgelert H5 chondrite that landed on the roof of a hotel in 1949.

- The Planetarium at the Royal Observatory, Greenwich has some meteorites on display, including a large iron that can be handled.

- The Armagh Planetarium (of which Sir Patrick Moore was once Director!) holds a small representative collection, including a whopping 140kg Campo del Cielo. (Stated to be the largest meteorite in Ireland!)

There are some much larger collections on display elsewhere around the world: if your travels take you near one of the following, it would certainly be worth making a visit.

- The Smithsonian Museum, Washington has a vast collection of meteorites, widely considered to be the largest and most important in the world.

- The Naturhistorisches Museum (Natural History Museum) of Vienna has one of the largest meteorite collections in the world: 7000 items representing 2400 different meteorites, over 1000 of which are on display.

- The Museum fűr Naturkunde, Berlin is another old collection, comprising 6000 specimens of 4100 different meteorites. This includes numerous thin sections which are also on permanent display and fine examples of Ensisheim, Krasnojarsk, and Nakhla.

- The National Institute of Polar Research, Tokyo has a huge collection – possibly the second largest in the world. It includes many specimens from Antarctica, among which are lots of rare achondrites and carbonaceous meteorites.

Some of the largest collections in existence are housed in academic institutions, many of which, sadly, do not permit visits by the general public:

- The Field Museum in Chicago claims to have the largest non-governmental archive in the world, largely based upon the purchase or bequest of a number of old meteorite collections. Although some material is made available on loan for outreach programs, the majority is reserved for research purposes.

- The Catholic Church owns another vast collection, which is curated by scientist-clerics of the Vatican Observatory at Castel Gondolfo, the Pope's summer home. Some of the meteorites have been used in spectroscopic and other tests and have made significant contributions towards our understanding of the bodies of origin of a number of meteorites.

What made all those holes?

AS mentioned earlierIt is a fact that every solid body within the Solar System is pock-marked by high-speed impact craters: not just the four rocky planets Mercury, Venus, Earth and Mars and their moons, but also the satellites of the four gas giants Jupiter, Saturn Uranus & Neptune. In addition, all the large asteroids that have been imaged from Earth or from space probes show similar dramatic evidence of past multiple impacts.

Currently, the most commonly-held belief is that the objects that created these structures were either large meteorites or small asteroids (accepting the definition that a meteorite is a chunk of metallic or rocky debris left over from the planetary formation phase or from the disruption of a planetissimal or asteroid).

For some time I have been reflecting upon the fact that very few of the meteorites I sell are associated with a known crater: some of course are:

- Barringer/Canyon Diablo, USA

- Wolf Creek, Australia

- Gebel Kamil, Egypt

But the vast majority of meteorites of all major types are found lying on the Earth's surface: there are no major craters associated with the huge Campo del Cielo or Sikhote Alin falls.

Furthermore, although there are around 175 large impact craters on Earth, none of those over 20km in diameter have produced meteorite finds! Strange!

A few, such as Chesapeake Bay, Gilf-el-Kebir, Darwin and Ries, have impact glasses associated with them, but wouldn't you expect that an object large enough to create a hole 20km across would leave significant solid remains?

It has been estimated that the iron meteorite that created the Barringer Crater in Arizona had a mass of 63,000 tonnes: of this around 30 tonnes have been collected as large fragments (up to 639kg) while another 8,000 tonnes probably remain as fine particles in the crater walls and floor. Rounding this up, we can suggest that about 12% of the impactor survived. Given that the Barringer Crater is quite small (around 1.2 km wide) then shouldn't a really huge structure such as the **300** km Vredefort Crater have a vast strewn field of hundreds of thousands of tonnes of material scattered around it?

The usual reason given for the lack of such material is that the vast majority was vaporised on impact and dispersed over a wide area. But even if this were true, there should be **some** big lumps, albeit some distance from the crater. And have any nickel-iron deposits been found in association with a large astrobleme? (impact crater) Well just possibly yes!

In 2012, I suggested in a magazine article that the huge Sudbury Ring structure was created by the impact of an asteroidal core that left behind the enormous reservoir of metallic minerals that is a globally important source of copper, nickel and iron. I have given several lectures on this theme and was a little surprised by the semi-hostile response I received on occasions!

To me this seems an eminently likely scenario: it is the only way to explain such huge deposits of dense elements at or near the surface. (You may recall that it is axiomatic that, during the differentiation phase of planetary formation, such elements sink inwards to form planetary cores) Some geologists have suggested that plumes rising from the core could bring heavy elements to the surface, but there is *no credible mechanism for this!*

To sum up:

- The Earth, like all rocky objects in the Solar System, shows numbers of structures produced by massive impacts throughout its history.

- Only a very small number of these craters have meteoritic material associated with them.

- There are just a couple of enormous astroblemes which are surrounded by rich deposits of metallic ores and minerals.

- The vast majority of craters must therefore have been produced by something other than meteoritic or asteroidal impacts.

In 1908 a massive explosion devastated around thirty square miles of Siberian pine forest near the Tungus River. Thousands of trees were flattened around an obvious epicentre and many animals (and quite probably local tribesmen) were incinerated or killed by the blast. Strange night-glows in the sky persisted for several days, while witnesses reported a blinding light and shattering concussion. However, despite numerous expeditions to the region, no satisfactory explanation has been forthcoming.

A 2014 TV documentary followed the adventures of a group of international scientists as they investigated the Tunguska impact site in Siberia: each one had a pet theory:

- An asteroid or meteoric impact

- An encounter with a chunk of anti-matter

- The explosive release of trapped gases from the mantle

- Aliens sacrificing their spacecraft to save humanity (No: really!)

You might well find it curious that these highly-qualified science professionals held such a diverse range of theories to explain a single event! The explanation for this is surprisingly straight forward: there is absolutely no concrete evidence at the site to prove conclusively what caused the explosion!

Apart from the felled trees, the only tangible 'evidence'

was found by a geophysicist who located eventually a small outcrop of sandstone: samples from this appeared to contain shocked quartz grains which, bizarrely, he claimed to be proof that an explosive release of mantle gases from beneath the Siberian Traps basalt layer had been responsible for the devastation! As many of you will know, shocked quartz is frequently encountered at the site of an extraterrestrial impact: in fact (along with shatter cones) such shocked grains are considered to be the 'smoking gun' of an impact crater.

Plate 21: Suevite from the Ries Crater

Little need be said about the anti-matter and UFO theories: there is, as yet, no proof that anti-matter exists, except as necessary extra mass to provide the gravity needed to hold the Universe together! And no fragments of exotic or alien technology have been located in the region of the explosion.

The generally-held belief is that Tunguska – and all large crater-forming events – were the result of an asteroidal impact. Unfortunately, no proven meteoric material has ever been found at the site: no nickel-iron, nor chondrite fragments. Neither have they been found at any other major impact craters such as Tenoumer, Manicouagan and Roter Kamm. This is strange, since many small to medium-sized craters are sources of stony or iron meteorites: Wolf Creek, Barringer, Gebel Kamil and Henbury being good examples. Even the giant Chicxulub event left behind only micro-tektites and raised iridium levels around the globe.

So what *did* explode above Siberia in 1908? And what made all the really large craters on Earth (and many of the small ones too!)?

In February, 2013 (thanks to phone and dashboard cameras) we were treated to grandstand views of an event that could easily have been a 'second Tunguska': a large object entered the atmosphere above Siberia before exploding over the city of Chelyabinsk.

As soon as the images of the meteor trail appeared online and on TV, I was struck by the fact that it was pure white, rather than the more frequently observed smoky grey-brown. Soon, reports from Siberia revealed that, despite intensive searching, only small, rounded fully-crusted wholestones were being recovered. This was in sharp contrast to the estimated entry mass of over 10,000 tonnes. Furthermore, according to a local meteoriticist, nothing extraterrestrial was discovered beneath the much-photographed circular hole in Lake Cherbarkul, merely an ancient algae-covered rock.

So what was the object that did all this damage? There could be only one explanation: a small cometary fragment.

Despite popular impressions, comets are neither rare, nor harmless cosmic snowballs: out at the very edge of the Solar System, the Oort Cloud and Kuyper Belt hold trillions of them.

For reasons that are not fully understood, one will occasionally tumble out of its orbit, plunging down the gravitational slope towards the Sun. Some swing round the Sun on a parabolic path and return to the depths of space: others are captured and remain in elliptical orbits, gradually sublimating as they pass by the Sun. Although the greater part of a comet's mass is water ice, it should be remembered that a cubic metre of ice has a mass of nearly a tonne: even a smallish comet 500m across, with a mass of around 1.04 million tonnes and a velocity approaching 70km/sec, would build up a devastating amount of kinetic energy as it accelerated into the inner Solar System.

As comets hurtle through space their surfaces become coated with a regolith of the material they have encountered during their passage. After a number of

'Sun-grazing' orbits of the Solar System, a comet's ice core will sublimate, leaving behind the regolith. If the debris should cross the Earth's orbit, this will form the raw material for periodic meteor showers.

It is entirely possible that collisions with small comets are responsible for those problematic large craters with little or no solid evidence of the impactor: much of the stony surface crust would be vaporised, producing a 'spike' of Iridium in the rocks, but no significant debris field (as would remain following an asteroidal impact.)

This hypothesis is gradually becoming more widely accepted, but remains contentious. Could it be that the money and resources spent searching for the small number of potentially Earth-crossing asteroids would be better spent elsewhere? Remember, aperiodic comets can appear at any time, with insufficient warning to mount any form of 'Bruce Willis' style mission!

A brief pre-history of man and meteorites!

THAT rocks and lumps of iron occasionally fall to Earth from the heavens has been known for thousands of years. In the past, when light pollution was non-existent and people lived much closer to nature, the fall of a meteorite must have been a spectacular and memorable event. Evidence of this is the fact that meteoric material has been found buried in ritualistic ways in several parts of the world. Strangely, the dogma of the Christian Church during the dark ages suppressed this, and it was not until the witnessed fall of the Wold Cottage L6 chondrite in 1795 that the extraterrestrial origin of meteorites was generally accepted.

There is a wealth of evidence that meteorites were considered sacred or special objects by our ancestors: here are some examples:

- In 1974 a walnut-sized H5 chondrite was discovered in a grain pit at the Iron Age Danebury Camp. Dating of the meteorite (using the Carbon 14 weathering method) suggests it was buried after it was seen falling nearby.

- At Winona, Arizona, a 24kg primitive achondrite was discovered wrapped in a feather cloth in a Sinagua burial cyst. This ritual interment mimics the burial of a child, indicating that it was regarded with reverence and respect.

- The Black Stone of the Ka'aba in Makka is generally considered to be a meteorite or possibly an impact glass from the Wabar Crater. Most Muslims believe it was given to Ishmael by the angel Jibreel (Gabriel) and it occupies a prominent location at one corner of the Ka'aba. Believers may touch or kiss the stone as they process around the structure during **Hajj**.

- A small Pallasite was discovered in a burial jar at Pojoaque, New Mexico: beads made of meteoric iron have frequently been discovered in Native American burial sites, notably in the Hopewell burial mounds.

- The Casa Grande iron meteorite has a mass of over 1,500kg. It was found inside a brick sepulchre in the ruins of a pre-Columbian temple: it was reverentially wrapped in linen in the same way as mummies from the same culture.

- Several important Greek and Roman shrine / temple complexes were built as repositories of meteorites: these were considered to be gifts from the gods. Well-known examples include the Palladion of Troy, the Cone of Elagabalus in Emesa and the sacred stone at the Temple of Artemis at Ephesus. The Temple of the Oracle at Delphi also contained a meteorite, believed by the Ancient Greeks to have been thrown to Earth by Cronus, the father of Zeus: it marked the Omphalus, or 'navel of the World'. There are a fair number of Roman coins that carry images of these sacred meteorites, indicating their importance in Roman mythology and belief. When the Syrian Elagabalus seized the emperorship of Rome he brought with him to the capital of the empire the Sacred Stone of Emesa: a black meteorite. This was housed in a new temple on the Palatine Hill, much to the disgust of the population.

From these and other examples, it would seem that meteorites have long been held in esteem as 'gifts from the gods': they were also used, more pragmatically, as a raw material.

Given that around 300 tonnes of meteorites fall across the surface of the Earth every day, it is not surprising that the 8% or so of these that are composed of nickel iron have been used by peoples of every inhabited continent. The tradition of making ritual halal knives from meteoric iron apparently began early in the history of Islam, and continues today. This may possibly be linked to the meteoric origin of the sacred Black Stone within the Ka'aba, or (more prosaically) to the requirement for a fine, razor-sharp blade without nicks or serrations.

It is widely known that Deputy Fuhrer of the Third Reich, Heinrich Himmler, had a fascination bordering on obsession for ancient sacred or ritual objects. He mandated expeditions throughout occupied Europe, Asia and Africa that searched with varying degrees of success for relics such as the Ark of the Covenant, the Spear of Destiny and the Grail Chalice.

In 1938, an expedition to Tibet led by zoologist Ernst Schäfer discovered a 10kg statue of the Buddhist god Vaisravana. Since the iron statue features a swastika as a central feature of its design, it was taken back to Germany and deposited in a private collection in Munich. In 2009 the statue was sold at auction and subsequently examined, when it was discovered to have been chiselled from a chunk of the Chinga ataxite that fell in the border regions between Mongolia and Siberia about 15,000 years ago.

It has long been established that meteoric iron was used extensively by the Egyptians and their neighbours the Beja (or Kushites) in the making of ceremonial and religious objects such as flails, maces and crooks.

The Old, Middle and New Kingdom Egyptians, who constructed the pyramids and most of the country's well-known monolithic structures, were effectively a Bronze Age culture: iron smelting did not take place in Egypt until around 800 BC. (The successes in war of the Hittites from the north was largely as a result of their secret knowledge of the processes of iron smelting) Any reference to iron before that time therefore refers to metal obtained from iron meteorites. In fact, the Egyptian word for iron 'bja' is best translated as 'the seed of the gods from heaven': this explains the origin of the pictogram for iron and also reveals that the extraterrestrial origins ofmeteoric iron were well-understood.

It is interesting to reflect that the Bedouins (who, with the Berbers, are among the most productive collectors of North African meteorites and impactites) possibly derive their name from the word 'bja'!

Of special interest is the fact that eighteen objects found in the tomb of the pharaoh Tutankhamen were made from meteoric iron: a model head rest, a knife blade and sixteen small ritual chisels.

Meteoric iron has been used to make both ritual and utilitarian objects by many ancient cultures: in North America both the tribal native Americans of the US & Canada and the Eskimo peoples of Arctic America and Greenland tipped spears and harpoons with slivers of meteoric iron. During his polar expeditions in the 1890s, Admiral Robert Peary was shown three large meteorites used for this purpose: he also discovered that iron from the Cape York meteorite in Greenland was traded widely by the Eskimo.

In conclusion it would seem that from the very earliest times the 'heavenly' origins of meteorites gave them a special significance to people who saw them descend: the fact that nickel-irons can be cold-hammered and chiselled made them a valuable source of metal among cultures that lacked the ability to produce the high temperatures needed to smelt iron ore. Interestingly, as recently as the Chinese Cultural Revolution of the 1960s, large amounts of the Nantan meteorite were sent to Beijing to be turned into iron!

Tektites and impactites

IN my humble opinion, no meteorite collection is complete without a few of these fascinating – even contentious – objects. Ranging from semi-precious gemstones to fossils contorted by the unimaginable energies released during an asteroidal or cometary impact, a representative collection can still be put together for just a few pounds.

Impact glasses are, as the name implies, transparent (or occasionally opaque, foamy materials) formed by the melting of sand and other minerals at the site of an impact or aerial explosion. (I have several pieces of **Trinitite**, a similar greenish-grey glass that formed beneath the nuclear test explosions in New Mexico in the 1940s and 50s.) There are literally dozens of known impact glasses (including the controversial gold-coloured Walton-on-the-Naze microtektites) but I'll restrict myself to a brief discussion of the more widely available or more interesting examples.

The best-known of the group is **Moldavite**, a beautiful apple-green glass that has been collected in the Czech Republic (and occasionally in Austria) for literally thousands of years, being valued for its colour, hardness and, in many cases, strangely beautiful shape and texture.

It is generally agreed that moldavite was formed by the event that excavated the Ries Crater in Western Bavaria, Germany around 14.5 million years ago. The crater is the largest so far known in Europe, measuring twenty four kilometres in diameter: the Bavarian town of Nördlingen grew up inside the crater walls! Ries is a classic contender for an origin as a cometary astrobleme, since no remnants of the object responsible for its formation have been discovered. Two vast aerial explosions carved

Plate 22: Shocked belemnites

out this and the nearby 4.8km Steinheim craters, shattering and recementing the local limestones into a fall-back breccia, known as suevite. (Enclosed in local limestones can sometimes be found fossil belemnites that were broken into fragments and then melted back together: fascinating evidence of the energies liberated by such events.)

Moldavite is a true form of glass, with the composition $SiO_2(+Al_2O_3)$ It cannot, therefore, be an altered limestone and must have originated as flints or erratic quartz within the local strata. The green colouration is considered to be due to the presence of iron *ions*: the iron, however, is not extraterrestrial and is probably from sulphide minerals such as pyrite and marcasite within the limestone.

The crater-forming explosion threw the molten moldavite high and to the south-east, the majority of it landing in the Moldau Valley, Czech Republic: genuine moldavite, therefore *must* be from this region (or very rarely, Austria, Germany or Poland) and items described online as African or Chinese Moldavite are fake.

Plate 23: Moldavites, suevite and fladles

Quarries in the Ries crater region also produce curious grey glass 'pancakes' (in German 'fladles') These are much larger than Moldavites and completely opaque: nevertheless, they are authentic silica-rich impact glasses that were thrown high into the atmosphere by the crater-forming event.

The only other genuine transparent impact glass comes from the Gilf-el-Kebir region near the Egyptian-Libyan border in the eastern Sahara and is variously known as Saharan, Egyptian, or **Libyan Desert Glass** (LDG). This semi-precious material ranges in colour from cream, through gold and bottle-green to smoky greys and browns. It is found in weathered fragments (some very large) in a hotly-contested and potentially dangerous

Plate 24: Genuine Libyan Desert Glass

region of the Sahara, and given also the demand for it within the jewellery and crystal therapy trades, it is becoming increasingly expensive and difficult to source. Yet again this has resulted in a flood of fake material in a range of bizarre colours: don't be tempted by 'Red Desert Glass' or 'Golden Tibetan Tektite': they only exist in the minds of Chinese fraudsters!

As with moldavite, LDG was formed by an aerial explosion, which, in this case, excavated a large, shallow depression, the Kebira Crater. This event occurred around twenty six million years ago, before the Mediterranean Basin flooded. At this time the Sahara was still a region of lush forest and grassland: a large population of Neolithic nomads lived here, who used LDG in the same way as the peoples of northern Europe used flint: to make tools and weapons. In fact, almost all the small pieces of LDG one encounters are scrapers or arrow points!

Inclusions such as cristobalite, shocked and vaporized quartz and traces of metals such as iridium, prove conclusively that the Kebira is an impact structure. (For example, Cristobalite, a form of silica, is only formed at extremely high temperatures.)

For thousands of years this beautiful glass has been valued for its subtle colours, waxy lustre and mysterious origin: it was prized by the ancient Egyptians, the most famous example of its use being the fabulous **Vulture Pectoral** found in the tomb of Tutankhamun by archaeologist Howard Carter. The Kheper Scarab at the centre of this jewel is carved from a glorious golden-coloured piece of desert glass.

Plate 25: Tutankhamun's pectoral (Courtesy Cairo Museum)

All other meteoritic glasses that are occasionally offered for sale are opaque and generally less attractive. Nevertheless, they are rare, very collectable and command increasingly high prices.

Darwin Glass is named after Mount Darwin, near Queenstown, Tasmania, as is the nearby ancient impact crater from which it is thought to have originated about 815,000 years ago.

Two varieties of Darwin glass are sometimes available: one is greyish-green while the other is generally black or very dark green. The two glasses have different chemical compositions, the dark form having less silica and more iron, nickel, magnesium, chromium, and cobalt than the paler glass. It's considered possible that the first form contains more material from the original impactor – although no fragments of this have so far been recovered. Australia rigidly controls the export of all meteorites and impactites: this, and the remote, snake-infested location of the crater, has led to the somewhat inflated price of Darwin Glass.

A very similar glassy impactite has its origins in the Sahara: **Aouelloul Glass** is a fascinating substance found close to a small (500m) crater in Mauretania. Unlike the majority of tektites, but like some Darwin Glass, this material contains traces of nickel iron from the meteorite that produced it.

The triplet of **Wabar** Craters in Saudi Arabia was formed by the impact of an iron meteorite with a mass of over two and a half tonnes. (Again, we see a large meteorite mass associated with very small impact craters!) Soon after their discovery in 1863, a small number of blackish impact glass spheroids was found nearby. These are sometimes referred to as 'Wabar Pearls' and are both highly-prized and very expensive! A frothy grey and white impactite similar to Aouelloul Glass also occurs near the craters.

Irghizite impact glass from the Zhamanshin impact crater, Kazakhstan is a black, twisted, rope-like material that may have formed during an impact that took place as recently as one hundred thousand years ago. Some researchers have suggested that this event may also have been the source of the tektites of the vast Australasian strewn field. (See below)

Two further, possibly twinned impact sites in Germany are those at Nalbach, Saarland and Chiemgau, near Munich. Both have produced Wabar-like black, contorted glasses which have been shown to contain the character-

Plate 26: Lonar Glass

istic cristobalite silica and both have a number of other unusual mineral and fall-back breccias in common. The suspected Saarland glasses are also found with a spectacular sky-blue colour. There are no obvious craters at either site (which are 500km apart) but this is possibly due to more recent alluvial deposition.

The **Lonar** Crater in India has produced several claimed impact glasses, one of which resembles a blue-grey amorphous silica, similar in appearance to Libyan Desert Glass, while the majority of specimens are froths and glass-rich breccias. The samples in my collection are very similar in density and texture to LDG, but remain very controversial.

Moving on, we now arrive at one of the most contentious groups in meteoritics: the Tektites!

True **Tektites** are chunks of glassy material found in a roughly equatorial belt in a number of locations around the globe. Most are blackish in colour and show signs of orientation and exposure to great heat. There has been much controversy about the genesis of tektites: their appearance and distribution in strewn fields has in the past been seen as evidence of an extraterrestrial origin on the Moon, but their chemistry is seen by many modern authorities as being more consistent with a formation on Earth.

It is generally believed that tektites were created when massive asteroidal or cometary impacts launched altered, molten surface rocks high into the atmosphere: remelting on re-entry produced the three orientations (shield, cone and sphere) characteristic of high speed ablation.

Plate27: Three orientations of Tektites

Other types, such as the large, stratified Muong Nong tektites, may have been formed on the Earth's surface by the energy of an impact, or of an aerial explosion (as was the case with moldavites and LDGs). The chief problem with this comparatively recent explanation is that no crater or surface feature has so far been conclusively shown to be the origin of the literally millions of Tektites that occur on Earth.

The majority of tektites offered for sale originate in what is generally called either the **Australasian** or **Indochinese** strewn field. Given the obvious physical differences between the tektites from various locations within these regions, it seems appropriate to consider each separately.

Australian tektites (Australites) are both unusual in appearance and very expensive! To me, some resemble liquorice Pontefract Cakes, being circular, flattened and with a raised rim: others are often described as boat-shaped. They have been found in ancient Native Australian campsites and middens, indicating that the Aboriginal population found them as intriguing as we do! Their existence became known to European scientists after Charles Darwin brought some back from his round-the-World trip in the mid-nineteenth century: he included an illustrated account of them in his 1844 publication *'Geological Observations on the Volcanic Islands Visited during the Voyage of H.M.S. Beagle'*, erroneously concluding that they were volcanic in origin.

Australites have been found pretty much all over the country, with perhaps slightly more being collected in the south. They generally seem to have been brought to the surface from underlying deposits on dry plains such as the Nullarbor by the action of the wind or of burrowing animals like Wombats, and have even been discovered in the crops of birds such as the Emu and Australian Bustard!

The chemical composition of Australites shows some variation between the population of Western Australia and the Nullarbor and those recovered from Victoria: however, they basically all consist of around 70% SiO_2 with varying amounts of metal oxides, including those of iron, sodium, aluminium, calcium and potassium.

There has been considerable discussion about where and when the Australites were formed: potassium / argon dating gives a far earlier age of 700,000 – 850,000 years, while carbon dating of organic matter buried with them gives estimates between 5,000 and 6,000 years.

Similarly, there is no real consensus about the location of the event that produced the Australites: some writers have suggested a common genesis for the Australites and the Indochinites somewhere in Laos, Cambodia or Thailand: others have plumped for a separate origin in north west Australia or even Wilkes Land, Antarctica!

Indochinites are found in the Philippines, Java, Malaysia, Indonesia and throughout the whole of the Indochinese Peninsula. They generally occur in the same range of ablated forms (disc, cone and sphere) although there is some geographical clustering of sizes, the largest seeming to originate in the Philippines and Vietnam. Those from the Philippines are called **Rizalites,** after a national hero, or (in the case of the particularly well-formed 'bread crusted' examples from the Biko Peninsula) Bikolites.

Bikolites in particular have a high-gloss surface with deep 'U'-shaped grooves running across their surface. The origin of these features is contentious, but they may have been formed during cooling or water erosion. (I have to say: large spherical examples are definitely my favourite Tektites – they just look so extraterrestrial!)

There are three broad divisions of the Indochinites (and some of the other types, too) based on their physical appearance. The majority occur in apparently ablated forms resembling orientated meteorites, with glassy interiors and smooth outer surfaces. Others resemble blobs of material that have impacted the ground while still molten: these are known as **splash form** tektites. Finally, **Muong Nong** tektites are found as irregular stratified chunks, often containing inclusions of the local rock.

For many years an extraterrestrial origin was considered highly possible for at least the first two of these. In the nineteen sixties, Geologists John O'Keefe and Hal Povenmire postulated that comparatively recent volcanic eruptions on the Moon could have blasted molten material into cislunar space, where it formed a thin Saturn-like ring system orbiting the Earth: this gradually fell to the ground, creating the observed equatorial distribution of the tektite strewn fields. (It has also been suggested that the impact events that created 'young' craters such as Tycho and Copernicus might have been an alternative source of lunar ejecta) Despite certain objections to an origin on the Moon based upon the chemistry of terrestrial tektites, it is absolutely the case that material collected on the Apollo 12 and 14 missions (A12014 and A14425) is chemically indistinguishable from the Indochinites.

Plate 28: Muong Nong, Rizalites and ablated Indochinites

Similar – but smaller – tektites are found in Africa and America: these, however, are subtly different in appearance and texture and are generally associated with known – or suspected – impact craters.

Ivory Coast tektites are thought to have been created during the excavation of the Bosumtwi crater in Ghana, which has been shown to be of approximately the same age (1.3 million years). They physically resemble Indochinites, but have somewhat different chemistries, containing less SiO_2 (67%) These are probably the rarest and most sought-after Tektites, not the least because of the volatile politics of the Ivory Coast.

Tektites originating in North and South America are usually referred to as **Americanites** (sometimes Amerikanites!). Again, their origin is controversial, with at least some examples certainly being obsidian (volcanic glass). Apart from chemical analysis (which can be inconclusive) a test that is sometimes used to identify genuine tektites involves heating strongly with a small blow-torch. Volcanic glasses such as obsidian tend to froth or foam: this isn't the case with genuine tektites. Some of the Americanites pass this test, others don't! Confusingly, some that *do* pass are found in eroded volcanic deposits: the jury is still out on these!

Texas Bediates and **Georgiaites** are both definitely true tektites, with all the right physical and chemical properties. Never very large (a 50g specimen would be exceptional) it is generally agreed that both originated in yet another 'undiscovered crater', believed to lie beneath

Chesapeake Bay. The tektites are found at the surface, having been eroded from Eocene deposits: as a result of their thirty five million year fossilisation, most examples have lost their original 'skin': those that haven't display furrows and 'navels' similar to Rizalites and are highly collected (and priced!)

Arizonaite / Saffordite is one of a number of apparent tektite types that were accepted as such for many years: apart from their lilac-grey colouration, they resemble the genuine article in most respects. Together with the similar **Healdsburgites** from California, however, the entire group is now regarded as volcanic in origin.

Colombianites bear a strong resemblance to authentic tektites, having textured surfaces and the general look of oriented and ablated glasses. Found in the Columbian jungle (often by local people searching for emeralds) Colombianites have an attractive lilac colour. Like Arizonaites, chemical analysis has confirmed them to be volcanic in origin.

A final group of authentic impact glasses are the microtektites. Found in numerous locations around the globe, these are all associated with major impacts.

Of interest to UK meteorite collectors and students are the tiny (<1mm) green, glassy glauconite spherules found in the **Bristol Impact Layer**. It is believed that this fine-grained Mercia Marl was formed from ejecta from the impact that created the 100km diameter Manicouagan

Crater in Quebec, Canada. First identified In 1973 at Churchwood Quarry, Wickar, Bristol, the deposit has now been completely destroyed by quarrying and samples are only occasionally available from the sale of old collections At the time the layer was deposited, Bristol was around 2000km from the impact site, since at that time the North Atlantic Ocean had not yet spread by tectonic activity.

Similar spherules are associated with the K-T impact (also increasingly referred to as the Cretaceous-Palaeogene event) that was responsible for the extinction of three quarters of all life on Earth 65 million years ago. Samples originating in the K-T boundary layer from the strata at Hell Creek, Wyoming contain numerous black microtektite spheres as well as large stressed quartz grains. (I've occasionally even found tiny reptile and amphibian bones in my samples!) The very largest K-T spheroids are found closer to the impact site (on the Yucatan Pensinsula) particularly on the island of Hispaniola. Some from the Dominican Republic can be 2-3mm in diameter.

As mentioned earlier, some intriguing microtektites have been recovered from the Red Crag deposits at Walton-on-the-Naze in Essex. Although considered controversial by some, I reckon these golden, glassy drops look the part!

Without doubt, impact glasses and tektites make fascinating additions to any collection: at the time of writing, it is still possible to obtain decent examples of most types at a fraction of the cost of a 'type set' of meteorites. This is also true of the final group in this section: crater impactites.

In a previous chapter, brief reference was made to the evidence used by geologists and meteoriticists when confirming that a crater or similar structure was formed by the explosion or impact of an extraterrestrial object. The energy released by the impact of even a small asteroid or comet is colossal: The Canyon Diablo impactor was probably about 30 metres in diameter, with a mass of 60,000 tonnes. About 30,000 kg of meteoritic material reached the ground at a velocity of more than 64,000 kph (including the 630kg fragment at the Meteor Crater Museum) The resulting explosion was equal to a **10 megaton** nuclear detonation, producing a crater measuring around 400 metres in diameter. Unsurprisingly, such an event will not only form a crater (which may disappear quite rapidly because of the forces of eroision) but will also cause characteristic alteration of the surrounding rocks.

At the time of writing, there are examples of over fifty impact breccias available online, originating from perhaps two dozen different sites. These include *suevites* from the Ries and Steinheim Craters in Germany to *pseudotachylites* from the 30km Rochechouart Crater in France. When cut and polished, many of these breccias are extremely attractive, often including fragments of impact glass and shocked minerals. My personal favourites are the various 'onaping' types and the carbonaceous anthraxolite from the Sudbury Crater in Ontario: not just

Plate 29: Sudbury impactites

fascinating to examine, but also a tangible link to one of the greatest explosions in the history of the Earth!

The pressure waves generated by a large impact travelling through surrounding rock layers almost always leave their mark in the form of **shatter cones**. Often said to resemble the imprint of a horse's tail, a large selection is generally available from specialist dealers or online: I have examples from twenty or so different impact sites in my collection.

For the sake of completeness, I should mention a branch of the collecting hobby that is, perhaps, more popular in the United States than the UK.

Very occasionally a meteorite will strike a man-made

Plate 30: Shatter cones

object such as a house, industrial building or even a car. In the US, fragments of these 'hammer stones' (as they are often called) are highly prized: there are even a couple of dealers who specialise in them. The demand is so great that even pieces of the building or vehicle are frequently offered for sale as 'authentic and historical souvenirs'!

I have to confess, I once received a consignment of Chelyabinsk 'peas' that also contained glass fragments from the windows of a local school, shattered by the original airburst: I included a piece with each of the little stones I sold!

A well-known example from this curious branch of meteoritics is the Peekskill fall of October 9th, 1992.

Thousands of people across New York State (including crowds at a number of football games) watched a bright green fireball plunging earthwards. A single large fragment of the meteorite responsible was recovered from the badly-damaged car of a teenage girl. Both pieces of the meteorite and of *the car* are frequently offered for sale!

So you think you've found a meteorite!

STATISTICALLY, the response is: no you haven't! For over twenty years I've travelled around the UK, selling meteorites and lecturing on astronomical topics. Perhaps not surprisingly, I have been shown literally hundreds of objects that the hopeful owners firmly believed to be meteorites.

Additionally, not a day goes by without someone 'phoning or e-mailing for the same reason. Unlike the majority of meteoriticists, I'm intrigued by these claims and still optimistic that one day someone will show me the genuine article, so that I always take the time to respond to **polite** requests. Sadly, only once has someone put in front of Linda and me an object they'd found in this country that was indisputably a meteorite!

Here are a few examples: some you might find amusing, some may cause you to roll your eyes: some may even produce a flash of anger! This is the spectrum of emotions that meteorite dealers experience on a regular basis! (To spare their blushes, I've removed any details that might help identify the people involved in these examples.)

A couple of years ago I was phoned by a builder who was renovating a large Georgian house on the south coast. He claimed to have found a large, spherical stone meteorite while widening the entrance to the building. I asked him if he'd tried any of the usual tests (see below!) and he confirmed that the object was strongly attractive to a fridge magnet. He asked what a large British meteorite would be worth: I of course replied that his find almost certainly **wasn't** from space, but that if it turned out to be an addition to the very small 'UK Club' its value could be in the tens of thousands.

I invited the finder to send me a couple of photos as e-mail attachments, which he duly did: after a brief inspection, I replied, explaining at some length how the object really didn't resemble any meteorite I'd ever seen. There was no response, so I assumed that was the end of the matter. I was wrong!

Two days later a 'jiffy bag' arrived: it contained a one-inch fragment of the putative meteorite, with a brief note from the builder saying that he'd broken it off the main mass with a 'Kango' hammer. A five second examination was enough to suggest that this small chip was, in reality, a lump of concrete: a couple of simple tests confirmed the identity. I left the sample in my workshop and e-mailed the owner with the good news!

Imagine my surprise when a few days later I received a letter on headed notepaper from the man's solicitor! I was astonished to read the following (paraphrased) threat:

"Dear Mr Bryant,

My client recently entrusted you with a piece of an extremely valuable meteorite. Unless you either return this or remit the sum of £1000 to cover its monetary value, my client has instructed me to commence legal proceedings."

To say I was astonished is understating things! I e-mailed the finder for clarification: he somewhat sheepishly admitted that he had just dropped the object, breaking it open to reveal a rusty sphere coated in concrete! I suggested he look on the other side of the gateway for its twin: this he subsequently did. As I had suspected from the start (the find having been made near a famous old naval base) the 'meteorite' was in reality a gate-post finial made from a cannon ball!

Om another occasion a local radio station 'phoned me for my opinion about a meteorite that a man in northern England had reputedly found in his garden that morning. The reporter told me that the stone was obviously genuine, '... because it was still hot enough to light a cigarette on!' As many of you may know, a meteoroid enters the atmosphere at high speed and at **very** low temperature (not much above absolute zero) It reaches the ground very rapidly, much of its super-heated leading face sloughing off, dissipating most of its initial kinetic energy. The interior and rear of the object is still ice cold! I already assumed that the new find was a hoax: the photos I received by e-mail revealed it to be a piece of red hot boiler slag that had presumably been blasted with a blow torch!

Another man had allegedly been struck by a 'stone from space' while watching a sporting event: a glance at the jpeg I was sent showed it to be a piece of red brick that, I assume had either fallen from the undercarriage of an aircraft or had been thrown from the crowd for some reason.

Now there is a curious fact about genuine stone meteorites: they are rarely as attractive as members of the public imagine they should be! Generally, a freshly-fallen stone will have a uniform, sooty outer surface, while an older one found by chance is usually indistinguishable from a somewhat dense beach pebble. In consideration of this disparity between expectation and reality, the first question I ask someone who thinks they've found a meteorite is "Is it attractive?" If the answer is yes, you can almost bet they haven't!

There is a semi-humorous term used by meteoriticists for the things that the public most frequently mistake for meteorites: ***meteorwrongs***! Here are a few classic examples:

The vast majority of the putative meteorite finds I am shown turn out to be one of the following: I have put them in order of frequency!

- **Marcasite nodules.** Marcasite is a variety of iron sulphide, FeS_2, which occurs quite commonly in limestone and chalk areas. It formed when hydrogen sulphide released by decaying organic material reacted with iron ions in ancient seas

Plate 31: Meteorwrongs

to produce an insoluble mineral. This is not ferromagnetic and doesn't test positive for nickel. Most nodules display beautiful radiating silvery crystals when broken or cut and, with their golden, lumpy surface, look like people imagine meteorites should!

- **Furnace slag.** Most people would be surprised to learn that, in times gone by, iron smelting was carried out in even the most isolated communities. Once the skill had been brought to the British Isles by the Celts, everyone wanted iron weapons and tools: it was like the Cold War weapons race 2800 years ago! The smelting process used charcoal, limestone and iron ore, which left a variety of residues. In addition, large amounts of this dense waste product were used as ballast in ships and to construct railway roadbeds. Almost any river or seaside foreshore in a ship building region will turn up literally tons of the stuff. This is what amateur meteorite hunters frequently send me for analysis. Some of these are dark, glassy, dense and may even be attracted to a magnet. The complete absence of **chondrules** and the frequent inclusion of bubbles and vesicles are good discriminators.

- **Belemnites.** These are the fossil remains of numerous species of prehistoric squid-like creatures that became extinct at the end of the Cretaceous Period. They resemble stone bullets with radiating glassy crystals exposed at the blunt end. They look like what people imagine something from space should look like, and were, in fact, referred to as 'Thunderstones' in the middle ages.

- **Haematite.** Many naturally occurring minerals and metal ores resemble meteorites in some way: mammiform haematite, with its curious, glossy surface, is something I am often shown by a hopeful finder.

- **Magnetite.** This is ferromagnetic and dense, and can develop a dark 'rind'. It is a very convincing meteorwrong, but, of course, tests negative for nickel.

All of the above conform to many people's preconceived ideas of what something from space should look like: here's the reality!

Plate 32: Selection of genuine untreated meteorites

So: Let's imagine that on one of your regular walks you come across an odd-looking stone that is completely different to those you normally encounter. Are there any tests you can carry out to eliminate the mundane?

- **Is it attracted to a magnet?**
  The vast majority of stony and iron meteorites are strongly attracted to a decent magnet because of the nickel-iron they contain. Only the very rarest achondrites are not.

- **Does it weigh more than it looks as if it should?**

  For the same reason as above, meteorites are generally surprisingly dense. Common chondrites are usually in the density range of 3.0 – 3.7 g/cm^3.

- **Does it have a fusion crust?**

  A meteorite acquires a thin, black crust because of the high temperatures generated by friction as it passes through the atmosphere. This will fairly quickly disappear on exposure to wind and rain, but a fresh fall should have a sooty, matt surface.

- **Does it display flowlines and/or regmaglypts?**

  The frictional forces mentioned above often sculpt the surface of a meteorite, producing thumbprint-like depressions called regmaglypts. If the meteorite has orientated itself in flight, it might also show shallow flowlines where molten rock has streamed away from the hottest front surface.

- **Are there small spherical shapes poking out of its surface?**

  Many (but not, of course, all) stony meteorites display chondrules on the surface. Beware, however of oolites and similar minerals which can look uncannily similar. There are even some permineralised types that have the right density and are even sometimes attracted to a magnet. (See the meteorwrongs above!)

- **Does it contain nickel?** Although nickel-containing minerals are not rare on Earth, most rocks wouldn't give a positive response to a nickel test. If you want to test your own finds, you can buy a test kit for less than £10 at high street chemist shops.

- **Does it have bubbles like an 'Aero' chocolate bar?**

  If so, even if it passes most of the tests above, it's almost certainly boiler or foundry slag. Very, very few meteorites are ***vesiculated***.

The ***single*** genuine meteorite find I've been shown? Many years ago I was invited to talk at the Institute of Astronomy in Cambridge. Afterwards a member of the audience approached me and, holding out his hand, said "I think I've found a meteorite!" Expecting the usual belemnite or marcasite nodule, I glanced down and was astonished to see what was immediately recognisable as a beautiful, orientated iron meteorite!

The finder, it transpired, was an archaeologist who had found the iron while field-walking in the Norfolk Brecks, As soon as I saw it, I was absolutely certain it was the real thing: I asked the finder if I could borrow it and take it to the Natural History Museum. He refused. I asked if I could do an on-the-spot nickel test. He refused. I asked if I could at least photograph it: you've guessed the answer already!

Apparently the owner was worried that his find would be taken away and become the subject of a coroner's treasure trove inquest, as is the case with archaeological finds. I was able to reassure him that this would not be the case, and that – provided he had the permission of the landowner to search where he had made his find – the meteorite would remain his property. It made no difference: the archaeologist thanked me, turned and left, he and his meteorite never to be seen again!

Supposing *I* wanted to find a meteorite in the UK – where would I start my search?

- **Dry stone walls**
  Many of these familiar features of the British countryside are far older than you might guess: the oldest date back to the Iron Age, while many in Yorkshire and the Peak District are over 500 years old! The farmers who built them as field boundaries or to protect their flocks picked up the material from the enclosed land. (This not only saved them a trip to the local garden centre, but also made ploughing easier) Any meteorites that might have fallen – and, don't forget, 300 tonnes land somewhere every day – will have ended up on the wall. The UK's only pallasite, Hambledon, was found in circumstances that suggest exactly this chain of events. It even had the mark of a plough-share across its surface!

- **Glacial moraine**

  This might seem a strange suggestion for a UK
  meteorite hunt, but, of course, much of the
  country was covered by ice sheets up to 300m
  thick during the five 'Ice Ages' that are known to
  have occurred in the past. Some of these have
  been so devastating that the entire planet was
  iced over: others – including the Quaternary Ice
  Age in which we still find ourselves – have been
  less severe. During an Ice Age, it seems that
  polar ice sheets periodically advance and retreat
  for reasons not fully understood: at the moment
  they are retreating, leaving behind deposits of
  sand and gravel and the piles of debris they
  collected. There are several types of this moraine,
  but of particular interest to us might be **ground
  moraine** and **lateral moraine**, examples of both
  being found in the UK.

  Ground moraine forms beneath a glacier and
  includes material washed under it by melt
  streams, as well as whatever was lying on top
  of it when it melted. Lateral moraine consists of
  parallel ridges of debris that either fell from the
  valley walls alongside the glacier as it advanced
  or from outer space, as has been found to be
  the case in Antarctica! In either event, moraine
  left in glaciated regions of the UK would seem to
  have the potential to produce concentrations of
  meteoric material in the same way as have the
  Allan Hills Glaciers in Antarctica.

- **The site of a previous known fall.**
  This is a fairly obvious suggestion: but searching a known strewn field always has the potential to produce further fragments. Of course, it's most likely that these would be quite small, but a search of unconsidered locations such as flat roofs, gutters and tree crowns could theoretically turn up a bigger chunk! Even grassland that has been extensively searched is worth a look, particularly if you have a metal detector: a small stony meteorite fragment is easily overlooked.

- **Prehistoric settlements**
  A bit of a long shot, this! However, as we've discussed above, our ancestors occasionally saw meteorites fall and generally seem to have treated any they found with a degree of reverence. The Danebury Camp meteorite was found in an iron age grain store and has been dated to the time of its construction. Of course, it is possible that the meteorite fell unnoticed into the store, but I reckon it's more likely that it was placed there for some ritual reason we can't even guess at!

Now members of the public can't just go poking around in archaeological sites, but if you're involved in an organised dig it would certainly be worth keeping your eyes open for rusty-looking bits of rock! It's one of those ironic facts that the more the human race learns, the more specialised it becomes: I seriously doubt whether the majority of recently-graduated archaeologists would recognise a meteorite (apart from the Cambridge field-walker, of course!) and I'd bet a few genuine examples have been thrown onto the spoil tips at digs.

To conclude: if you are really determined to find your own meteorite in the British Isles, you might well have a long search ahead of you! However, if you happen to travel widely around the globe, any arid region (whether hot desert or frozen ice field) is worth a look: just don't forget that that the rarest and most interesting meteorites are **not** attracted to a magnet!

Meteorites mentioned in the text

QUITE a few passing references have been made in the text to historically or scientifically important meteorites: more details about any of these can, of course, be found online. If you're reading this in the bath or bed, though, the following brief descriptions might be of use!

- **Barwell**

  One of less than fifty meteorites discovered in the UK, this L5/6 stone fell in the village of Barwell on Christmas Eve, 1965. Careful searching at the time (and since!) has produced around 40kg of material. The original mass is generally referred to as being 'Turkey-sized'!

- **NWA 869**

  This huge strewn field was first discovered near Tindouf, Algeria in 2000. Classified as an L5/6 chondrite, it displays fusion crust, metal flecks and brecciation, as well as occasionally obvious chondrules. At least some of the items offered for sale are possibly from other falls in the same region, resulting in occasionally different lithologies.

- **Parnallee**

  This impressive 77kg LL3.6 was seen falling in India in 1857. When cut and polished it displays large, beautiful dove-grey chondrules in a dark matrix.

- **Murchison**

  With a main mass of around 100kg, this is one of the larger carbonaceous meteorites known. It was seen falling in several pieces in 1969 and classified in the CM group, having a fine-grained matrix, smallish chondrules containing hydrated minerals and various other inclusions.

- **Tagish Lake**

  In January, 2000, following a loud explosion, a fireball was seen descending onto the frozen surface of this Canadian lake. Around 10kg of fragments were discovered before (it is frequently claimed) the lake melted, causing the remainder to be lost forever!

  The matrix of this ungrouped carbonaceous stone contains amino acids and primitive pre-solar grains.

- **Allende**

  Often considered the most-studied of all meteorites, this CV3 carbonaceous chondrite fell on 8th February, 1969 in Chihuahua, Mexico

- **Tatahouine**

  This cumulate diogenite was seen to fall near the village of Tatahouine, Tunisia, on June 27th, 1931. This fascinating stone is generally agreed to have originated beneath the surface of the asteroid Vesta. Tatahouine was, of course, chosen as the name of Luke Skywalker's home planet in the Star Wars films!

- **Almahatta Sitta**

  The small asteroid 2008 TC3 was tracked during its approach to Earth and observed during its descent both from space and the ground. It landed on October 7th, 2008, near Station Six (Almahatta Sitta in Arabic!) in the Sudanese Nubian Desert. It was identified as being a Ureilite and contains numerous nano-diamonds that make it arduous to cut and polish!

- **Al Haggounia 001**

  This three tonne fossil meteorite was found in Morocco in 2006, initially being classified as an enstatite. Having been the subject of much debate, it is now accepted as an Aubrite.

- **NWA 2999**

  Around 400g of this dark-surfaced achondrite were purchased in 2004 by an American dealer, the twelve stones having probably been found in Algeria. Classified as an Angrite, some authorities consider it displays characteristics of an origin on the planet Mercury.

- **Shergotty**

  One of the original three classifications of Martian meteorite is named after this 5kg stone that was seen falling in India in 1865.

- **Nakhla**

  The second of the 'snick' Martians is named after the Egyptian town where this 10kg stone fell in 1911. There is a popular belief that a fragment of the stone struck and killed a local dog, an event commemorated in the company logo of an American dealer!

- **Chassigny**

  This, the rarest of the three SNC Martians, was seen falling in France in 1815: at the time of writing only three Chassignites have been identified. The stone had a mass of 4kg.

- **Vaca Muerta**

  This famous mesosiderite was found in 1861 in the Atacama Desert in Chile. The Spanish name translates as 'Dead Cow', a reference to a nearby gully into which cattle presumably wandered occasionally! The total mass of the many fragments collected is just under 4 tonnes.

- **Bondoc**

  This attractive meteorite was found on the Bondoc Peninsula, Luzon Island, the Philippines in 1956: it was sent to legendary meteoriticist Harvey Nininger, who identified it as a mesosiderite.

- **Campo del Cielo**
The most popular 'entry level' iron meteorite for serious collectors! First recorded by Spanish *conquistadores* in 1576, over 50 tonnes of fragments of this famous find have been located so far, mostly in regions of Argentina that have preserved them in excellent condition.

- **Sikhote Alin**
No collection should be without a chunk of this, the *'meteorite that nearly started World War 3'* Exploding in the skies near the Soviet naval base in Vladivostok, it is reported that local authorities were convinced it was a nuclear attack. Fortunately, this was 1947 and Russia lacked the capability to respond...
Examples are found as pieces of twisted, shiny 'shrapnel' or as much rarer and more expensive oriented individuals.

- **Hoba**
The 60 tonne Hoba 1VB iron was found in 1920 by a Namibian farmer. This is the largest known single meteoric mass and, for this reason, remains exactly where it was discovered!

- **Chinga**
In 1913, just under 210kg of metal fragments were found by gold prospectors near the Chinge River in Russia. It is believed that the original ataxite meteorite exploded above a glacier some 10,000-20,000 years ago.

- **Gibeon**.
  Classified as a fine octahedrite with a typically narrow Widmanstätten bandwidth, this iron meteorite was discovered in Namibia in 1838. A number of large examples may be seen mounted on plinths in the city of Windhoek, the capital of Namibia.

- **Millbillillie**
  This curiously-named eucrite was seen falling in Western Australia in October, 1960. Searches commencing around eleven years later recovered a total of 330kg, the largest fragment having a mass of 20kg.

- **Middlesbrough**
  The beautiful, oriented 1600g Middlesbrough L6 chondrite was seen falling in 1881. It displays exquisite flowlines and fusion crust and is, in my humble opinion, by far the most handsome meteorite yet discovered in the UK.

- **Canyon Diablo**
  Undoubtedly the most famous meteorite crater in the world is the Barringer Crater, Arizona. The original huge coarse octahedrite mass fell approximately 30,000 yrs ago, about 4 miles east of the Canyon after which the meteorite is named. Iron from the crater region was collected and used by Native Americans and a mining company was once established to exploit the presumed but undiscovered main mass: much of this is now generally believed to have vaporised on impact.

- **Wold Cottage**
  This famous L6 meteorite was seen falling in
  Yorkshire on December 13th, 1795 and, at 25kg,
  remains the second largest observed fall in
  Europe. It played an important role in the debate
  about the extraterrestrial origin of meteorites and
  is commemorated by a brick-built obelisk at the
  site of the fall.

Known or suspected bodies of origin of some meteorites

UNTIL comparatively recently it was generally assumed that most stony meteorites were ancient, accreted pieces of the solar nebula. It was theorised that irons and pallasites were the debris from a small number of collisions between differentiated planetissimals, as were the few known achondrites. As was discussed earlier, it was as recently as 1982 that it was first suggested that the Allan Hills 81105 meteorite from Antarctica was lunar in origin. After careful examination, it was realised that a previous find – Yamato 791197 – also had its origin on the Moon: shortly afterwards the first Martian meteorites were identified.

Increasingly refined techniques in chemistry, spectroscopy and physics, (as well as greater levels of general acceptance!) have allowed meteoriticists and exogeologists to assign bodies of origin to many meteorites, frequently with high levels of certainty.

Asteroidal meteorites

The fall of asteroid fragment **2003 TC** in October 2008 provided an indisputable link between an asteroid and several classes of meteorite (in this case ureilite and bencubbinite).

Before this event, however, the technique of **relative reflectance spectroscopy** (RES) had demonstrated that it is possible to assign a specific asteroidal origin to a wide range of meteorite types. To understand how this works, a little review of high school science is needed!

White light is, as most people know, a bundle of different wavelengths of electromagnetic radiation that we call 'visible light'. Each of these wavelengths is interpreted by the brain as a different colour, which is known as the **spectrum of visible light**: in order, from long to short wavelength, red, orange, yellow, green, blue, indigo and violet. Our eyes see an object because of the visible light it reflects: a white object reflects every wavelength, a black one reflects none, while a green object reflects green light and absorbs the wavelengths at the red end of the spectrum. (From this you can infer that most plants absorb and use red light for photosynthesis!)

A glass prism can split sunlight into the various wavelengths (as Sir Isaac Newton famously demonstrated) If you place the prism (or, to be honest, more usually a diffraction grating) at the end of a telescope, it can also produce the spectrum of a distant luminous object. Stellar astronomers have used this technique for over a century to investigate the motion and composition of stars and galaxies.

Many years ago it was realised that the light reaching

the Earth from the thousands of asteroids that orbit the Sun would vary, depending on which wavelengths had been reflected or absorbed by the body's surface layers. More recently it was discovered that the RES of an asteroid could occasionally be matched to that of a chunk of meteorite: the following graph shows how closely the RES of the Tatahouine diogenite matches that of the asteroid 4-Vesta.

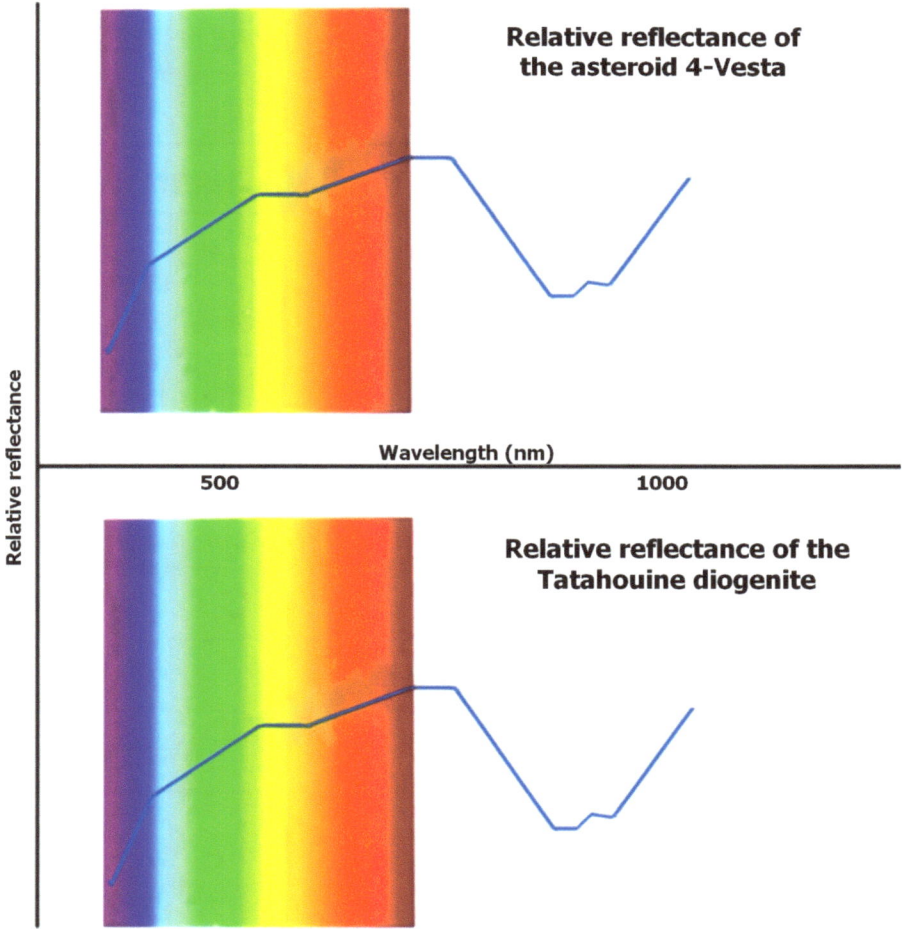

Plate 33: Relative reflectance spectrum

This process has made it possible to suggest a number of asteroids as origins for meteorites: here are a few examples:

Origin	Meteorite class
Nysa	Aubrite
Eger	Aubrite
Vesta	Howardite
Vesta	Eucrite
Vesta	Diogenite
Eros	L4
Psyche	E4
Ceres	CM2
Hebe	H
Boznemcová	LL / L
Pallas	CR
Nenetta	Brachinite
Asporina	Pallasite

Plate 34: Asteroid origins

Planetary meteorites

The history of the discovery of the SNC achondrites and subsequent additions to the Martian inventory has been described in a previous chapter: you'll remember that the chemistry of gaseous inclusions in these meteorites was found to match samples obtained during robotic exploration of Mars. It was also discovered that the relative amounts of two isotopes of oxygen (^{17}O and ^{18}O) found in the thin atmosphere of Mars is different from that on the Earth: it became apparent that meteorites could be grouped according to their **relative oxygen isotope concentrations** (ROICs).

However, the only other major planet on which robot probes have touched down has been Venus, and none of the Russian **Venera** landers that achieved this remarkable feat were specifically tasked to sample ROICs. In any case, those that *did* manage to reach the surface of Venus mostly survived less than an hour because of the planet's crushing atmospheric pressure and hostile environment.

Interestingly, High Resolution Spectroscopy and results from the Vega, Pioneer and Magellan probes suggest that the relative concentrations of Venus' oxygen (fixed in carbon dioxide molecules) is within that found in rocks from the Earth-Moon system.

Despite the present incredibly dense atmosphere of Venus, several mathematical models suggest that meteorites *could* have been launched during impacts on the planet before this evolved.

In 2012 a beautiful glassy, greenish-coloured meteorite (later named NWA 7325) was recovered from the Sahara 'somewhere in Morocco'. Analysis of a sample revealed that it had a similar composition to that dis-covered by the **Messenger** probe as it orbited the planet Mercury: substantial quantities of magnesium, aluminium and calcium silicates but very low amounts of iron. Furthermore, its ROIC indicates an origin on a differentiated body that does not lie close to Mars, the Moon or Earth on the distribution chart:

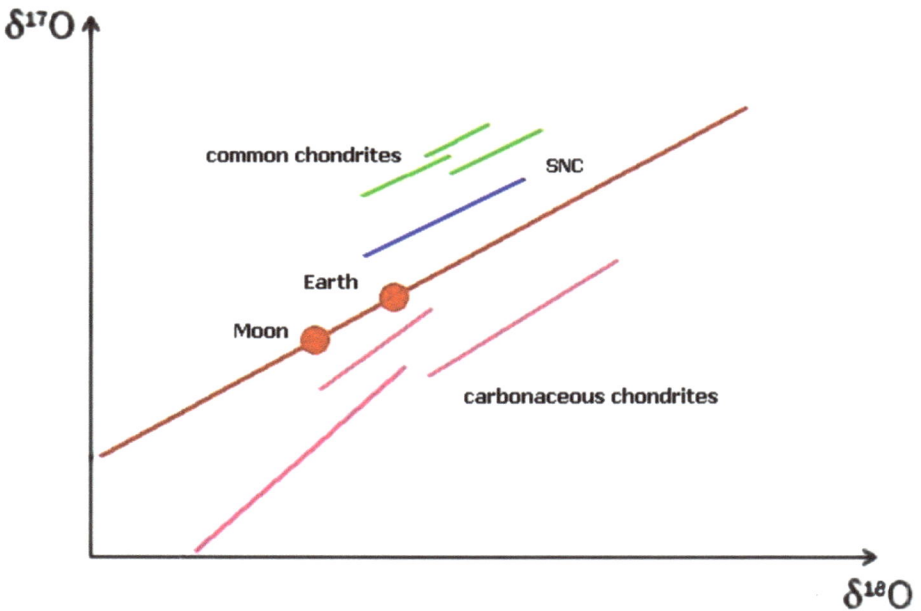

Plate 35: Relative oxygen isotope chart

At the time of writing, the jury is still out, some meteor-iticists suggesting that an asteroidal origin for NWA 7325 is more likely, with others pointing to its chemical composition as evidence that the achondrite is the first so far to have reached us from Mercury.

Nevertheless, analysing meteorites and clustering them by ROIC not only allows exogeologists to identify any that belong to a certain group, but also those that don't! A fascinating discovery has been that some enstatites and brachinites lie on the terrestrial fractionation line! As unlikely as it might at first seem, the many huge impacts the Earth has experienced since it formed could easily have launched material into space, particularly before our dense, oxygen and water vapour-rich atmosphere formed some two and a half billion years ago.

Carbonados

These black diamonds are known from two strewn fields on either side of the Atlantic: Brazil and the Central African Republic. Both regions lack formations containing Kimberlite (the usual source-mineral of diamonds sold in the gem trade) which has caused many authorities to suggest an extraterrestrial origin. Their age (over three billion years) and location indicate that they reached the Earth before the Atlantic Basin opened as a result of tectonic plate movement, possibly as the result of a single vast impact.

Other possible origins that have been suggested include material projected from a nearby population 1 or 2 stellar supernova and the spontaneous fission of thorium and thorium within the Earth's crust: if either were the case, you'd expect to find a more widespread distribution.

More interestingly, some exogeologists have put forward the theory that carbonados may have been formed by pressure during the collision between a very large asteroid and one of the Solar System's gas giants. The most likely of these would be the seventh planet, Uranus. This strange world has its axis of rotation in the same plane as its orbit and effectively 'rolls' around the Sun on its side. Many possible explanations have been proposed for this eccentricity, but a massive collision would seem the most likely. Since the atmosphere of Uranus contains a high percentage of hydrocarbons such as methane and ethane (as well as some carbon monoxide) the energy of a planetary impact might easily compress these molecules into diamond, just as is thought to have been the case with stars such as the white dwarf BPM 37093.

Plate 36: Carbonados

Planetary satellites

Apart from Mercury and Venus, every planet in the Solar System has one or more natural satellites. (Our Moon is the largest relative to its primary and should more correctly be considered a twin planet of the Earth).

Each of these satellites is pock-marked with craters left by cometary or asteroidal impacts, many of which must have projected debris into space: you might think that these would have very similar ROICs to their primaries, making identification problematical. However, since most planetary satellites are actually captured asteroids or planetissimals, the majority appear to have entirely different compositions. Apart from the lunaites, just a single meteorite has so far been suggested as possibly having an origin on a planetary satellite: the unusual Kaidun CR2 carbonaceous chondrite that fell in the Yemen in 1980. The inclusion of alkaline-rich clasts from a fully-differentiated planet within a carbonaceous matrix indicates an origin on a small world very close to a larger one, and Kaidun's chemistry suggests Phobos is a strong possibility.

There must be very many more meteorites with an origin on a planetary satellite orbiting out there: the challenge is to recognise them and identify their primary!

Price guide

THE meteorite market is so volatile that it would be pointless to publish a list of current retail prices: it would be out of date before this book reached the printers!

However, the price of a meteorite – like anything else – is linked closely to the demand for it and to its availability. As an example: when the first Chelyabinsk cometary fragments came on the market in 2013, even small examples a centimetre in diameter fetched £80+ Once the initial interest died down (and as more were discovered) the price tumbled: the same stones were offered for sale at £10 or less. After a couple of years, the supply began to dwindle and prices started to creep up again!

Generally speaking, historically or scientifically significant meteorites command the highest prices, followed by attractive or impressive types. Stones are the most abundant, followed by irons, stony irons and, finally, achondrites: this is reflected in the financial value of each group.

Another factor that influences the price of a meteorite is its appearance: a rusty chunk of Sikhote Alin is worth a small fraction of the value of a carefully-curated, beautifully oriented example.

Finally, location is often a telling factor in value: most collectors want to own a sample from their own country and in some cases there just isn't enough to go around: a 1mm x 1mm fragment of the famous UK Barwell fall, for example, would currently fetch at least £50!

To give an idea of what a collector might have to pay to own some typical examples of each type, here's a list using a 'star' system that runs from one star (abundant and relatively inexpensive) to five stars (rare, beautiful and highly sought-after).

As a rough guide: a 5cm weathered common chondrite from the Sahara might be available at around £5 – £10. For the same amount, you could probably buy a 15mm fragment of the CV3 NWA 3118: both are thus awarded a single star in the table!

Stony meteorite price guide

					meteorite	size	category
★					Unclassified, weathered,	50mm	common chondrite
★	★				Unclassified, fresh	50mm	common chondrite
★	★				Classified, abundant	30mm	common chondrite
★	★	★			Unclassified, oriented	30mm	common chondrite
★	★	★	★		Classified, abundant, oriented	30mm	common chondrite
★	★	★	★		Classified, scarce	30mm	common chondrite
★	★	★	★	★	Classified, scarce, oriented	50mm	common chondrite
★					Classified small fragment	15mm	carbonaceous chondrite
★	★	½			Classified small wholestone	15mm	carbonaceous chondrite
★	★	★			Unclassified polished slice	30mm	carbonaceous chondrite
★	★	★	½		Classified larger wholestone	30mm	carbonaceous chondrite
★	★	★	★		Classified polished slice	30mm	carbonaceous chondrite
★					Small eucrite fragment or slice	10mm	HED
★	★	½			Small eucrite wholestone	10mm	HED
★	★	★			Larger eucrite fragment or slice	30mm	HED
★	★	★	★		Small howardite fragment or slice	15mm	HED
★	★	★	★	½	Larger howardite fragment or slice	30mm	HED
★					Small abundant diogenite	5mm	HED
★	★	★			Larger abundant diogenite	10mm	HED
★	★	½			Small scarce diogenite	5mm	HED
★	★	★	★		Larger scarce diogenite fragment / slice	20mm	HED

Iron and stony iron meteorite price guide

					meteorite	size	category
★					Campo del Cielo crystal	25mm	iron
★	★				Sikhote Alin 'shrapnel' fragment	20mm	iron
★	★	½			Canyon Diablo fragment	30mm	iron
★	★	★			Muonionalusta, polished / etched slice	50mm	iron
★	★	★	½		Sikhote Alin, oriented, small	30mm	iron
★	★	★	★		Campo del Cielo, polished / etched slice	10cm	iron
★	★	★	★	★	Large Campo del Cielo (good condition)	1kg	iron
★	★				Gibeon, small fragment or etched slice	15mm	iron
★	★				Nantan, small weathered fragment	30mm	iron
★	★				Imilchil, small fragment	15mm	iron
★	★	★	½		Gebel Kamil, 'lizard skin' sample	60mm	iron
★	★	★	★		Sikhote Alin 'shrapnel' fragment	80mm	iron
★	★	½			Sericho, untreated whole sample	50mm	pallasite
★	★	★			Sericho, polished end cut	40mm	pallasite
★	★	★			Sericho, polished slice (double sided)	40mm	pallasite
★	★	★	★		Seymchan, polished slice (double sided)	50mm	pallasite
★	★	★	★	½	Pallasovka, polished slice (double sided)	50mm	pallasite
★	★	★	★	½	Seymchan, polished slice (double sided)	90mm	pallasite
★	★	★	★	★	Imilac, polished slice (double sided)	40mm	pallasite
★	★	½			Bondoc, polished slice	30mm	stony iron
★	★	★			Vaca Muerta, fragment	40mm	stony iron
★	★	★	★		Gujba, polished slice	30mm	stony iron

Tektite and impactite price guide

				meteorite	size	category
⯪				Indochinite fragment	30mm	tektite
★				Indochinite, complete	50mm	tektite
★	★			Indochinite, complete	100g+	tektite
★	★	⯪		Muong-nong fragment	60mm	tektite
★	★	★	⯪	Rizalite, complete	30mm	tektite
★	★	★	★	Australite, button or 'boat'	30mm	tektite
★	★			Libyan Desert Glass fragment	40mm	impact glass
★	★	★		Libyan Desert Glass fragment	50mm	impact glass
★	★			Moldavite, fragment	30mm	impact glass
★	★	★		Moldavite, complete	30mm	impact glass
★	★	★		Bediasite / Georgiaite	30mm	Americanite
★				Shatter cone	50mm	impactite
★	★			Suevite, pseudotachylite or similar	100mm	impactite

Some books worth reading!

Tektites – a Cosmic Paradox
Hal Povenmire, 1997

Tektites in the Geological Record
Joe McCall, The Geological Society, 2001

Catalogue of Meteorites
Monica M Grady, Cambridge University Press, 2000

The Cambridge Encyclopaedia of Meteorites
O. Richard Norton, Cambridge University Press, 2002

Field Guide to Meteors and Meteorites
O. Richard Norton & Lawrence A Chitwood, Springer, 2008

Meteorites and their Parent Planets
Harry V. McSween, Jr., Cambridge University Press, 1987

Solar System Evolution: A New Perspective
Stuart Ross Taylor, Cambridge University Press, 2008

Meteorites: A Journey Through Space and Time
Alex Bevan & John de Laeter, Smithsonian Institution Press, 2002

Glossary

I'VE tried to keep this book as easy to read as a page-turner novel, but inevitably a few technical terms were bound to sneak in. Most of these have been explained as they cropped up, but here are a few that might need a word or two:

Accretion: The cohesion of small particles under the influence of gravitation to form larger bodies: this is the method by which dust in the solar nebula gradually came together, eventually forming planetissimals and planets.

Asteroid: A rocky world, smaller than a planet, that formed by accretion within the solar nebula. Most occur in fixed orbits within the main belt between Mars and Jupiter, but a few are found elsewhere and roam dangerously close to the Earth at times!

Breccia: A rock (or meteorite) formed by the bonding together of fragments of other material, by heat, pressure or chemical action.

Differentiation: The gradual separation into layers of the various components of a planetary body, the denser elements sinking inwards under the influence of gravity to form a core.

Equilibration: The process by which the original primitive structure of meteorites (and the bodies from which they originated) is metamorphosised by chemical reactions, hydration, heat and other factors. Unequilibrated meteorites may display ancient features such as chondrules and pre-solar grains.

Escape velocity: The minimum velocity needed for an object to break away from the gravitational attraction of the body it is resting on. The more massive the body, the greater its escape velocity: that of the Earth is about 11.2km/s, while that of the Moon is just 2.38km/s.

EVA: A classic NASA acronym – 'extra-vehicular activity': it just means leaving your spacecraft.

Fossil meteorite: Meteorites have been crashing down on the Earth since its formation: a fair percentage must have landed in lakes, seas and deserts, where they were covered by sediments that eventually formed rocks. Occasionally these are discovered by quarrying or mining. Some examples include al Haggounia 001 and several examples from Thorsberg in Sweden.

Isotopes: Atoms of the 118 elements in the Periodic Table differ one from another in the number of protons in their nuclei (which is called the atomic number) This defines the element, and does not vary. However, atomic nuclei also contain neutrally-charged particles called neutrons: the number of which *is* variable. For example, a hydrogen atom can have 0, 1 or 2 neutrons (and even as many as 6 in some very short-lived forms!) Each configuration of the atom is called an isotope of that element.

Mafic/Ultramafic: Mafic rock types are silicate minerals or igneous rocks rich in magnesium and iron (for example, basalt). Ultramafic rocks, such as many lavas, peridotites and dunites, have a much lower silica component (less than 45%).

Orientation: As a meteorite enters the atmosphere it adopts a stable aerodynamic position (rather like a space-craft re-entry module) During the extreme heating and ablation that follows, the meteorite will often acquire one of three characteristic shapes: shield, cone or sphere. It may also develop **flow-lines** of molten material streaming away from its leading edges and **roll-over** lips of material flowing over the back surface.

Paired: Independent searchers may collect samples of new meteorites and send them off for classifi-cation to separate authorities. Each having been examined and given a name or number, it may later be decided that they originated as one fall. An interesting example is the pair of achondrites NWA 7325 and NWA 8014, which may have originated on the planet Mercury.

Thanks!

IN everything I write, do, or achieve, there standing at my shoulder with words of advice and encouragement is my beautiful, clever wife Linda. We started our **'Spacerocks UK'** business some twenty years ago and it says so much about Linda's determination and intelligence that – despite an academic background in languages – she achieved an OU degree in Science (including Astronomy) so that she could speak confidently with our clients (and keep a check on me!) **Sans toi, ma chérie, rien!**

Many thanks to my old friend Nik Szymonek for his most flattering foreword. Nik is, by common consent, among the top exponents in the world in his field of astro-astronomy. I am incredibly flattered that he took time out of his busy schedule to say such generous things about this book and its author.

One of my few regrets in life is that, as a blues singer and guitarist, I've never had the chance to 'jam' with Nik – in addition to his many other talents, he is a fabulous rock drummer!

Many thanks are due to the friends and colleagues throughout the world of meteoritics and astronomy who have helped make the last twenty years possible.

There are far too many to mention them all, but chief among them are Dr.Carolin Crawford at the Institute of Astronomy, Cambridge, the amazing German planetary meteoriticist, Stefan Ralew, our American friend Robert Cucchiara, and last, but never least, our thousands of clients, many of whom have become dear friends.

As always, thanks to my excellent designer-editor, Bob Tibbitts for his patience and the terrific advice he provides.

If you've enjoyed this book, you may like to read my other related volume:

DANGER
FROM THE SKIES
The real threat to mankind's existence!

DAVID BRYANT

www.ingramcontent.com/pod-product-compliance
Lightning Source LLC
Chambersburg PA
CBHW041307020426
42333CB00001B/7